HIDDEN ATTRACTION

HIDDEN ATTRACTION

The History and Mystery of Magnetism

GERRIT L. VERSCHUUR

New York   Oxford
OXFORD UNIVERSITY PRESS
1993

Oxford University Press

Oxford   New York   Toronto
Delhi   Bombay   Calcutta   Madras   Karachi
Kuala Lumpur   Singapore   Hong Kong   Tokyo
Nairobi   Dar es Salaam   Cape Town
Melbourne   Auckland   Madrid

and associated companies in
Berlin   Ibadan

Copyright © 1993 by Oxford University Press, Inc.

Published by Oxford University Press, Inc.,
200 Madison Avenue, New York, New York 10016

Library of Congress Cataloging-in-Publication Data
Verschuur, Gerrit L., 1937–
Hidden attraction : the mystery and history of magnetism /
by Gerrit L. Verschuur.
p.   cm.
Includes bibliographical references and index.
ISBN 0-19-506488-7
1. Magnetism—Popular works.
2. Magnetism—History—Popular works.
3. Physicists—Popular works.   I. Title.
QC753.5.V47   1993   538—dc20   92-37690

2   4   6   8   9   7   5   3   1

Printed in the United States of America
on acid-free paper

Preface

A glorious mix of drama, inspiration, insight, good and bad judgment, hard work, and sometimes sheer luck mark the path of scientific discovery. This is seldom more clearly illustrated than in the quest to understand one of nature's most remarkable phenomena, magnetism. The hidden source of its almost magical powers of attraction has been sought for over two thousand years. In this odyssey, creative human beings have moved from superstition to certainty and synthesis, from an era when answers to basic questions could only be imagined to the present time, in which scientists are able to incorporate magnetism into a great scheme of the basic forces in the universe. To see how progress was made is to watch the evolution of a science. And by recognizing how scientists moved from superstition to synthesis, we are reminded a little of ourselves, as we struggle to dig out from under the welter of beliefs and prejudices about the ways of the world and strive for a clearer understanding of the mystery of our existence. There is every indication that once our curiosity is aroused, and given enough time, the human intellect can and does discover the answers to its questions. How this progress was manifested in the case of magnetism will be the theme of *Hidden Attraction*.

The saga of discovery that led to the solution of the mystery of magnetism began over two thousand years ago, when explanations of natural phenomena could only take the form of beliefs rooted in fantasy. But then, after many centuries of blind acceptance of such beliefs, skeptical and curious individuals began to challenge dogma and superstition, and their questioning slowly and inevitably led to experimentation. That turned out to be the reliable way to get at the truth about the reality that underlies appearance. The transition to experiment marked the birth of the scientific era, about four hundred years ago.

I will argue that without the stunning progress made during the last several centuries in understanding the nature of magnetism, our modern technological civilization would not yet have come into existence. Every facet of the civilized world rests, ultimately, on the widespread availability of electricity to drive the machines of industry. We would never have learned to produce electricity if it were not for the profound insights that arose from the study of magnetism. As a result of ever-deeper probing into the nature of reality, we have even learned to reach out and sense magnetism between the stars.

Hidden Attraction concerns an adventure of the mind, and these chapters have been conceived to trigger your imagination and stimulate your curiosity. An adventure of discovery underlies all of science, something that is easily ignored, in large part because of the mushrooming welter of facts that overflow the time available for their communication. Those who make judgments about what should be taught to stay "up-to-date" tend to rule against the human and historical side of science. This is sad. As a result, the curricula of our schools lose touch with the dramatic and exciting roots of human thought.

Despite the presence of footnotes, which are offered for those who wish to explore further, *Hidden Attraction* is not meant as a definitive work of history. I have taken the liberty of touching on only those incidents that I felt were important to the story. Similarly, I have related something of the life of a few scientists whose work seemed essential to progress in the quest that forms our theme. In this context, I owe a debt of gratitude to the many historians of science whose research reports provided fertile territory into which I ventured to gather material for this book.

I am indebted to several people, some anonymous, who read early versions of this manuscript and made helpful suggestions. I am particularly grateful to Mike Town for his comments and encouragement, and for the careful reading and constructive criticism by a student of science, Kim Hill. The encouragement of my wife, Joan, and my editor, Jeffrey Robbins, is also deeply appreciated. Finally, I dedicate this work to those young minds who may be inspired by something they read here to pursue a career in scientific exploration.

Lakeland, Tenn.   G. L. V.
September 1992

Contents

HIDDEN ATTRACTION

n I n
Of Mystery
and Magnets

The Magnet's name the observing Grecians drew
From the Magnetic region where it grew
Lucretius, as quoted by William Gilbert, *De Magnete*

O NE of my earliest memories
is of vivid flashes of light-
ning followed by claps of thunder that filled me with terror. To
deal with the awesome light and noise I turned to my parents
for reassurance. I wanted an explanation. But how could my
parents deal with my tearful questions? Even if they had known
about convection, charge separation, electrical discharge, temporary vacuums, and shock waves, how could they relate such
facts to a child? I needed an immediate guarantee that thunder
and lightning were nothing to worry about.

In their attempt to calm me, my parents instinctively resorted to the time-honored technique of *inventing* an explanation to suit the moment. When in doubt, create. It doesn't matter whether the answer has a bearing on reality. For them it
was a case of any port in a thunderstorm. To help allay my
terror they offered me an answer that my child's brain could
understand. On the far side of the lake next to which we lived,
they said, by the larger lake that lay beyond the railroad tracks,
lived a man flashing a searchlight. This lit up the clouds and
created lightning. (It was wartime and I had seen a searchlight,
so this story made sense to me.) Someone else was upending a

dump truck loaded with empty forty-four-gallon oil drums, so they claimed, and that caused the thunder.

In retrospect it occurs to me that their explanation had the hallmarks of a superstition in the making. I was supposed to believe what they said. The dictionary definition of a superstition is a belief resulting from ignorance or fear of the unknown, or trust in magic or chance. I was certainly both ignorant and afraid of the unknown. Superstitions are further defined as beliefs that are a state or habit of mind in which trust or confidence is placed in some person or thing. I wonder whether my parents hoped that I would trust those invisible gentlemen engaged in the otherwise harmless practice of flashing searchlights and upending dump trucks in the middle of rainy nights. Whatever their rationale, it was only after tearful debate that I accepted their explanation, as a working hypothesis one might say.

When primitive human beings attempted to account for mysterious natural phenomena, their initial explanations could only be invented, usually through liberal applications of fantasy and imagination. There was no other way to come up with an answer, at least not until the invention of clever measuring devices extended human senses.

Ten thousand years ago it was *possible* to ask questions about the nature of the world, but it was impossible to find answers that pertained to the true nature of things. For example, it was possible to ask what a star was but impossible to know; at least until the invention of the telescope, in 1609, and, more particularly, spectrometers in the nineteenth century, devices for studying the detailed nature of light. It was also possible to ask what lightning was but impossible to know about the nature of violent electrical discharges. Only in the twentieth century would lightning begin to be understood, and even now the details continue to be argued. But it will not be the details that concern us in our quest to understand magnetism. Instead, I will show that after many centuries of persistent inquiry the broad picture is clear. The questions have been answered, and with remarkable consequences. Humankind has lived for a long time with questions about natural phenomena but only recently have we discovered ways to find answers that pertain to reality. Here I define reality as the way the universe is, rather than the way we might wish it to be. Experience often confirms that there is a large difference between these two points of view.

Answers to fundamental questions, when first formulated, relied on myth, superstition, and a vast array of less formalized but nevertheless wild beliefs. Beliefs begat further beliefs, usually building upon those that went before, all equally untested against "reality." As a result, no one ever got closer to the truth, at least not until someone challenged the old beliefs and began to confront nature directly, which meant through experiment. Herein lies the essence of the scientific endeavor, and of our story. Our shared quest for truth could move out of the realm of gods and powerful spirits, so often invoked in times of need, only with the discovery of the scientific method for studying nature: experimentation under controlled conditions and close attention to the information contained in our observations of either nature or our experiments.

If one pays attention to experiment and experience, authoritarian beliefs handed down through the generations often begin to feel hollow. This is surely to be expected, since, at root, beliefs are arbitrary. It is possible to believe in anything you want and no one can fault you for it. But whether a given belief pertains to the nature of reality is another matter entirely. In this rests the substance of the great divide between science and religion. While scientists may hold a belief about the nature of reality, in the form of something they call a hypothesis, such a belief is held only until new observations or experience reveal the hypothesis to have been only an approximation of the truth, if not actually unfounded. Then the hypothesis will be modified to take into account new information, or the new experience of one's senses. But a belief can be held quite independently of what an individual's experience may reveal. Beliefs are essentially subjective mental constructs and as such tend to act as powerful blinders to the facts of one's experience.[1] For example, it is possible to call it a miracle to survive a tornado that killed one's neighbors and destroyed their homes. But the scientific approach to why certain houses were devastated would depend on observation. It would not take long to discover that the path of a tornado depends on random factors, on chance— good or bad luck, depending on one's point of view—and no amount of belief in miracles could have changed its course.

Nevertheless, superstitions and beliefs have served individuals in useful and highly subjective ways for a long time, even if the beliefs ultimately have little or nothing to do with reality, at least as it pertains to the objective world of existence outside our minds. But once our beliefs begin to approach reality, they enter the realm of scientific hypothesis that can be tested through experiment. A hypothesis may deserve the label of theory if it can be used to predict further phenomena or events. In some cases, the study of nature, guided by theory, may even lead to the discovery of the laws that underlie the nature of reality.

Over the last two thousand years, critical minds have learned to confront and transcend ancient beliefs about natural phenomena. They found ways to learn about the true nature of those phenomena. That required experiment and paying heed to data obtained from observation. In the case of my initial belief about thunder and lightning, I became increasingly dissatisfied with the explanation. If there were men doing strange things on rainy nights, where did they live? Could I talk to them and ask them why they did it?

My quest took on an experimental form after I constructed a primitive canoe made of corrugated iron and blocks of wood, with liberal amounts of tar applied to cracks in the vessel to prevent its foundering. At first I paddled around in front of our house, within secure reach of the shore. Then, gradually, I confronted the fact that I could paddle over to Big Lake, which could only be reached by water. If I took the risk, I might find out where the men with the searchlight and the oil drums lived.

This adventure of discovery was no small undertaking for several reasons: I was not at all sure the makers of thunder and lightning would welcome my snooping, my canoe was highly unstable and required constant bailing, and Big Lake was far away, at least as seen from the perspective of a small boy. It was beyond the three islands of Small Lake, which meant paddling out of sight of home, a safe haven in case of trouble, and under the railroad bridge. Beyond that was the unknown.

One momentous day I undertook my voyage of discovery. I paddled along the railroad, beyond the islands, out of sight of home, and paused under the bridge. At the third or fourth attempt I ventured across the frontier. Small waves lapped dangerously at the bow of my canoe and threatened to engulf it.

Ready to retreat at the first sign of hostility from the "gods of thunder and lightning," I looked around. Weed-covered banks and reeds extended to the horizon. Not a trace of a searchlight. No houses where anyone with a searchlight could even have lived. Not even a pile of empty oil drums.

I was relieved. But I was propelled once again into the limbo of uncertainty. The only belief I had held up to then about thunder and lightning, the only straw I had clung to in order to alleviate my fear, had been destroyed. Curiosity and fear had to turn elsewhere for their satisfaction. In the end this required that I study physics and learn to observe lightning carefully so that I might discover its nature. To find the answer would take many years and in the end I relied mostly on what others had learned in similar quests.[2] Therefore, I was not moved to fly a kite in lightning storms, as Benjamin Franklin had done. In science it is possible to learn from the experience of others. Satisfied with what they had demonstrated to be true, I turned to other questions that intrigued me. One of those concerned magnetism.

Long-forgotten individuals began to ask questions about magnetism over two thousand years ago. Nature had provided an abundant supply of magnets, called lodestones, to pique their curiosity. Lodestone is a form of iron oxide called magnetite found in the shape of bands in certain iron mines. The origin of magnetite layers in these mines is itself remarkable, a story we will defer until Chapter 11. It will become apparent, however, that without an abundant supply of lodestone for inquiring minds to consider we would not now be enjoying the benefits of modern technology. That is not to claim that without lodestone to show the way humankind would never have invented electric generators, motors, radio, or television, but that such things would, at best, have come along much later than they did.

The name lodestone referred to the ability of small, elongated fragments of this metallic material to indicate direction. A simple compass needle made of lodestone could *lead* you to your destination in the same way that the lodestar (Polaris, the pole star, which marks north in the heavens) could guide you on a journey.

It was the obvious attractive powers on the lodestones that

set them apart from other natural phenomena. Their power was reminiscent of the ability of a piece of amber (fossilized tree resin), which, when rubbed, drew "feathers, straws, sticks and other small things"[3] toward it. But lodestone worked without anyone having to rub it. That made it far more magical.

What is immediately important is that lodestone is magnetic, a concept now familiar to almost everyone. It draws other pieces of iron to itself for no apparent reason. This magical behavior originally gave rise to explanations couched in terms of outright superstition. Just as my parents resorted to wild fantasy to account for a mysterious phenomenon, so early explanations for magnetism were pure fiction. And why not? How else could primitive minds cope with such an awesome mystery?

Nature's magnets have been known since hundreds of years before the Christian era. In the first century B.C. Lucretius wrote that "iron can be drawn by that stone which the Greeks call Magnet by its native name, because it has its origin in the hereditary bounds of Magnetes, the inhabitants of Magnesia in Thessaly [northern Greece]."[4] He also knew that magnets could repel. "Sometimes, too, iron draws back from this stone; for it is wont to flee from and follow it in turn."

An alternative theory about the origin of the word magnet was proposed by Pliny the Elder, who wrote that there was a shepherd named Magnes, "the nails of whose shoes and the tip of whose staff stuck fast in a magnetic field while he pastured his flocks."[5]

Over the centuries beliefs about the lodestone's powers grew. In the thirteenth century Bartholomew the Englishman hailed its medicinal properties in his encyclopedia:

> [T]his kind of stone [the magnet] restores husbands to wives and increases elegance and charm in speech. Moreover, along with honey, it cures dropsy, spleen, fox mange, and burn. . . . [W]hen placed on the head of a chaste woman [the magnet] causes its poisons to surround her immediately, [but] if she is an adultress she will instantly remove herself from bed for fear of an apparition.

He also claimed that "there are mountains of such stones and they attract and dissolve ships made of iron."[6] This prospect must have curdled the blood of many able seamen.

At about the same time, in 1269, the first attempts to choose truth over superstition regarding the phenomenon of magnetism took place on a battle-torn field outside the walls of the besieged city of Lucera in Italy. In those dark days, when men did little but fight other men for the right to control what people believed, Lucera was under siege for the third time in fourteen years. A crusade had been launched against its people, the Crusades being the polite name given to wars sanctioned by the papacy against anyone believed to be the enemy of christianity. As a reward for being involved in "God's work," the pope was wont to hand out the title Peregrinus (which means pilgrim) and thus Pierre de Maricourt, of unknown birth date, became Peter Peregrinus.

Peter Peregrinus was a member of the army of the King of Sicily, Charles of Anjou, and probably served as an engineer. During his service he wrote a letter, an *epistola*, that described everything known about lodestones and how to make instruments using these magnets. Although the bulk of the letter concerned his efforts to invent a perpetual motion machine using magnets,[7] its impact was vast. His letter was "completed in camp, at the siege of Lucera, in the year of our Lord 1269, eighth day of August"[8] and sent to his soldier friend, Sygerus of Foucaucourt. Apparently Peregrinus had shaped a piece of lodestone into a sphere, and in his letter he explained how parts of it acted on an iron needle. He discovered the existence of two magnetic poles and was the first to coin the term "polus" to refer to the north and south ends of a magnet. He performed simple tests with floating lodestones to show how they attracted or repelled one another, how they were drawn to point north–south, and how a lodestone could be used to magnetize an iron needle. All this occurred outside the walls of Lucera, where people were being starved to death to encourage them to re-think their beliefs. It was there that Peter, in his role of engineer, must have been performing some holy task such as solving problems related to battering down city walls.

This mysterious man, who was patiently serving his king while waiting for the citizens of Lucera to submit, was, according to Roger Bacon (c. 1220–c. 1292), one of the world's first experimental scientists. Bacon described Peter Peregrinus as a master of experiments and thus by experience he knows natural, medical, and alchemical things, as well as all things in the heavens and beneath them; indeed he is ashamed if any layman, or grandam, or soldier, or country bumpkin knows anything that he himself does not know.

. . . nothing is hidden from him which he ought to know, and he knows how to reprobate all things false and magical.[9]

Roger Bacon was himself a pioneer of experimental science, which, as a Franciscan friar, he taught at the universities of Oxford and Paris. I suspect he was also the chairman of the Peter Peregrinus fan club, at least if the following praise of the engineer is anything to go by:

Without him it is impossible that philosophy could be completed, or be treated usefully or with certainty.

For should he wish to stand well with kings and princes, he would find those who would honor him and enrich him. Or, if he were to show in Paris by his works of wisdom all that he knows, the whole world would follow him: yet because either way he would be hindered from the bulk of his experiments in which he most delights, so he neglects all honor and enrichment, the more since he might, whenever he wishes it, attain riches by his wisdom.[10]

The mystery about Bacon's adulation is that it was expressed before Peter even wrote his epistle about the magnet, the only document of his that has survived until modern times. One can only wonder what else he must have written about with such authority. It is even more bizarre that this person, whom, according to Bacon, the "whole world would follow," was engaged in laying siege to a town during the Crusades. How did he find time to perform his experiments (on the side as it were) while the stark reality of man's inhumanity to man was playing itself out all over the Italian countryside?

To return to Bacon's point about Peregrinus setting a great example, until that time the study of the natural world was totally determined by what one *believed* to be true. It was that type of thinking that launched the Crusades, given that they concern matters of faith. The beliefs in question were based on what others had originally written or preached a long time before. The greater the amount of time that had passed since the death of the authority figure the better. In fact, back then, when it came time to make sense of life or to express the urge to understand natural phenomena, experience or experiment counted for nothing. The notion of experiment had barely been considered. Those who were interested in the study of nature spent their time considering the writings of dead philosophers, such as Aristotle or Plato, rather than looking at the real world, at the nature of nature. That is like relying on weather reports on television to enjoy good weather rather than going outside and experiencing it directly.

Peter Peregrinus was apparently one of the first, if not the first, to stress that "experience rather than argument is the basis of certainty in science."[11] This is what made his epistle so important as a historical document. In that letter he applied his revolutionary idea to the study of the lodestone and stated his goal, which was to explain "how iron is held suspended in air by virtue of the stone." That desire set in motion a train of events, albeit a slow train, that seven hundred years later allows us to enjoy the benefits of electricity.

Above all, Peter Peregrinus asked why lodestones were imbued with the peculiar ability to attract one another over a distance. Until then all answers had been steeped in superstition, and he was aware enough to see that something was amiss. He began his inquiry by logically dispelling some widely held notions; for example, that lodestones pointed toward the mines where they were originally found. Since stones from different parts of the world all pointed north–south, they clearly ignored their home base. Their place of origin therefore had nothing to do with the phenomenon. His alternative explanation was more heavenly. The force that drew the lodestone had to be in the heavens, specifically that region directly above the earth's North Pole marked by a point around which the North Star, Polaris, draws a small circle in its nightly motion.[12] Therefore, because a magnet pointed north–south, it couldn't be Polaris itself that was attracting the lodestone. It had to be the invisible celestial pole. He therefore concluded that the poles of a magnetized needle, as well as every part of a spherical lodestone, received their power from the corresponding part of the celestial sphere (the imaginary globe against which our senses tell us the stars are projected).

If this concept were correct, he argued, one should also be able to observe a spherical magnet slowly rotating about its axis just as the heavens appear to rotate around the earth. After all, if the celestial sphere and all parts of it mysteriously imbued the magnet with its properties, then the magnet should be tied to the apparent motion of that sphere. Peter attributed his failure to detect such a rotation, that is, to make a successful test of this theory, to his lack of skill in building a suitable device rather than to any fault in the theory itself.

It has been said that "The Epistola [on the magnet] ranks as one of the most impressive scientific treatises of the Middle Ages."[13] Peter Peregrinus had brought together all that was known about magnets at the time and set the stage for the birth of the science of magnetism. However, 330 years were to pass before anyone else took up the task of moving the subject of magnetism's mystery out of the realm of superstition into that of science. In retrospect it could not have happened much earlier, partly because the idea of science was itself new, at least science as we understand the meaning of the word today.

Three centuries is a long time, especially when measured against the rapid rate of intellectual and technological change in our modern age. Yet it was not until 1600 that William Gilbert (see Chapter 2) would build upon the groundwork laid by Peregrinus by single-handedly confronting the widespread misconceptions and superstitions that still surrounded the nature of magnetism. Even then it was another century or so before other scientists began to grasp at the truth. As we shall see, Gilbert was also way ahead of his time. The scientific study of magnetism required more novel concepts than either Peregrinus or Gilbert could have imagined. The phenomenon would never be understood until it could be brought under control and magnetism created at will. Also, it would require the invention and development of measuring devices that permitted the study to be placed on a quantitative footing. These two requirements, being able to create more of what it is you wish to study and being able to measure how much of it you have, are essential for the scientific study of virtually any physical phenomenon.

In the interim, though, attempts to refine the manufacture of magnetic compasses were undertaken. Mariners had become aware of a peculiar deviation between true geographic north and the direction in which the compass needle pointed. Not only that, but this horizontal deviation—declination as it was called—varied from place to place on the earth's surface.[14] One could interpret this to mean that unless you had a map that showed the magnetic declination of a region the compass was not much use. If you were lost, a compass didn't help, unless you knew where you were so you could look up the corrections to the direction of north at that location—in which case you weren't lost!

This peculiar phenomenon of the variation, or declination, of the needle, produced one hugely significant historical consequence. It determined whether Christopher Columbus discovered America in October, 1492.

As with any major journey of exploration, Columbus took along a good lodestone, which was carefully guarded, and a supply of spare compass needles. These would be remagnetized with the lodestone if they lost their ability to seek north.

Columbus's ships sailed west by the compass, but that was not geographical west. During the voyage they had unknowingly sailed past the point where the deviation of the magnetic pole was zero; that is, where the magnetic and geographic poles were in the same direction. Beyond that point magnetic north lay to the west of true north instead of to the east, as was true back in Spain. This error caused his ships to head further south than would otherwise have been the case. This was important. Under pressure from his restless crew, who wanted to turn back, Columbus agreed to a specific time at which, in the absence of sighting land, they would return to Spain. Had they been sailing toward the geographical west that moment would have been reached well out of sight of land.[15] Instead, thanks to the magnetic compass, Columbus stumbled onto land and found fame.

The manufacture of compass needles was deemed to be a great skill in the fifteenth and sixteenth centuries. But all was not well, as the experience of instrument maker Robert Norman in 1581 illustrates. He told the story in his book *The New Attractive*,[16] delightfully subtitled, "Containing a short discourse of the magnet or lodestone, and amongst others his virtues, of a new discovered secret and subtle property concerning the Declination of the Needle, touched therewith under the plain of the Horizon. Now first found out by Robert Norman, Hydrographer."

In Norman's time there was much argument about the correct way to rub a compass needle and how the quality of the lodestone used affected the final product. Many practitioners of the art of making a good needle swore that the only reason it pointed away from true north was because the maker had rubbed it the wrong way.[17] For Norman the problem was more severe. Whenever he mounted a new magnetized needle it tilted with respect to the horizontal. To make it stay level, he added a small counterweight, even if this solution robbed him of the satisfaction of having made a perfect instrument. He suspected that his failure in making a perfectly balanced needle was due to shortcomings in his skill or imperfections in the raw material.

This balancing act was used year after year because no matter how carefully he rubbed the needle with the lodestone, the final product always tilted until he added a counterweight.

One day he had an inspiration. Instead of adding a weight at one end of the needle, why not cut a piece off the other end and see if it balanced? No. It still tilted. So he cut off some more. Still no good. He tried again and again, and whittled away his precious needle until there was nothing left! At this point Norman was literally forced into making his marvelous discovery. As we shall see repeatedly in our saga, nature's lessons are difficult to learn. Norman's experience was no exception. So it came to pass that in the year 1576 he made a particularly large and well-magnetized needle and found that it also tilted. Then it occurred to him to find out whether the tilt was real. To do so he set an unmagnetized needle to swing vertically. Good. It didn't tilt. Then he magnetized it and suspended it again. It came to rest at an angle to the horizontal!

At last Norman understood, and he was elated. Something was pulling on the needle to make it tilt. He had uncovered one of nature's secrets, but what did it mean? Was something pulling up on one end, or was it pulling the other end down? To find out he mounted a needle in a cork and fiddled with the size of the cork until the combination was neutrally buoyant (did not rise or sink in water) and experimented to see what happened. Under the water the needle again tilted, but the cork/needle combination did not rise to the surface. Neither did it sink to the bottom. That was very peculiar. He interpreted the observation to prove that *nothing* was pulling the magnetized needle up or down. He scratched his head and concluded that the attractive force acting on it was *nowhere*. He convinced himself that the power to tilt the needle had come from the original lodestone used to magnetize it. He called the tilt "The Line Respective" and confessed that he could explain its cause as well as he could account for the movement of the celestial sphere, which, he admitted, was not at all.

None of this stood in the way of his writing a book about his wondrous discovery. It was dedicated to "the right worshipful Master William Borough, Comptroller of her Majesties Navy," and Norman offered his secret humbly, hoping that it would benefit many. Because the secret was contrary to what had been believed up to then, Norman, who described himself as an unlearned mechanician, was more than a little worried about the consequences of publication. Revelations of nature's truths had not always met with applause. (That remains true to this day; witness the continuing reaction to the notion of evolution.) On the contrary, excommunication, torture, and general discomfort on fiery stakes were hazards faced by many of those whose thinking did not conform to established dogma.

Nevertheless, Norman was excited by his wonderful discovery and found solace in recalling that Archimedes, upon figuring out how to calculate the specific gravity of a gold crown, had leaped out of his bath naked and "came crying to the King his master, I have found it, I have found it!" But he, Norman, so he pointed out, was not moved to the same loss of self-control. On the contrary, upon making his wondrous discovery, he neglected his "own nakedness and want of furniture" to make it known to the world in the form of a book. In the dedication he added that "I thought it my duty to adventure my credit and make my name the object of slanderous and carping tongues, rather than such a secret should be concealed, and the use thereof unknown." [18] In other words, come what may, the world had to be informed.

> And seeing it has pleased God to make me the instrument to open this noble secret, that his name be glorified and the commodity of my country procured thereby. [19]

He could not hide the truth he had found, despite fearing for his life lest he disturb the authorities that controlled thought. He insisted that his discovery would benefit navigation and stressed that he based his arguments only on "experience, reason, and demonstration, which are the grounds of the Arts." Today we would say that these were the hallmark of good science. To me, Robert Norman was one of the world's first scientists.

He was so moved by the wonders of magnetism that he waxed poetic:

> Magnes, the lodestone I,
> your painted sheaths defy,
> without my help on Indians seas,
> the best of you might die.
>
> I guide the Pilots course,
> his helping hand I am,
> the mariner delights in me,
> so doeth the merchant man.
>
> My virtues are unknown,
> my secrets hidden are,
> by me the Court and Common weal,
> are pleasured very far.
>
> No ship could sail on seas,
> her course to run aright,
> nor compass show the ready way,
> were Magnes not of might.

Norman had recognized that the magnet looked toward what he called "the point respective" but was not *drawn* toward it, wherever that point was. If it had been, the magnet and cork combination placed under water should have drifted up or down. It was left to William Gilbert twenty-four years later (see next chapter) to recognize that the magnetized needle tilted with respect to the earth because the earth is magnetic, a brilliantly correct deduction. Today physicists appreciate that the forces acting on the north and south pole of the compass needle act only to align it. Since the pull of the two poles is equal and opposite, compass needles don't go flying off into space, or plunging to the ground. The same cannot be said for what Jonathan Swift's Gulliver found on his travels.

In the country of Balnibarbi, Gulliver found a floating island called Laputa located a little off shore. At the center of the island was a great lodestone six yards in length and three yards wide. It was firmly mounted and yet so well balanced that the weakest hand could turn it. This meant that the island could be made to move by rotating the lodestone to a suitable orientation: up, down, or sideways. All you had to do was twiddle the lodestone and the island would wander every which way. But when the lodestone was horizontal the forces up and down were equal and the island did not budge. Needless to say, he who controlled the lodestone controlled the island. That led to nasty intrigues and no one lived happily ever after.

Meanwhile, back on planet earth, and seen from the perspective of history, Peter Peregrinus was way ahead of his time. His rational approach to an otherwise magical phenomenon represented a radical change from the way other philosophers had confronted nature's mysteries until then. But few people took note, at least not until 1600. Only then would the first major step be taken in humankind's search for an understanding of magnetism.

NOTES

1. This has been wonderfully argued by John C. Lilly, *Simulations of God: The Science of Belief.* (New York: Simon & Schuster, 1975, and Bantam, 1976).

2. I wrote about the explanation, which remains for me the most

impressive, in an article entitled "Supernovae and Thunderstorms," *Astronomy,* May 1984, p. 48. But I am no expert in the subject and gather that arguments still rage as to which theory is correct.

3. William Barlow, *Magneticall Advertisements,* 1616. (Republished by Da Capo Press, Amsterdam and New York, 1968).

4. Quoted by William Gilbert, *De Magnete,* Book V, Chapter XII. (London, Chiswick Press, 1900).

5. *Encyclopaedia Brittannica,* 1970. Volume 14, p. 311.

6. Quoted from Christopher Cerf and Victor Navasky, *The Experts Speak.* (New York: Pantheon Books, 1984).

7. A highly readable summary of Peregrinus's letter is given by Park Benjamin in his book *A History of Electricity.* (New York: Arno Press, 1898).

8. Peter Peregrinus, *Dictionary of Scientific Biography.* (New York: Charles Scribner's Sons, 1970–1980).

9. Ibid.

10. Ibid.

11. Ibid.

12. Due to a slow, twenty-six thousand-year wobble of the earth's axis in space, called precession, the point in the heavens directly above the North Pole does not remain fixed but sweeps out a circle. The star Polaris just happens to be very close to that point now, but it isn't quite at what is called the north celestial pole. Thus Polaris, like all other stars, moves in a circle around the pole.

13. Peter Peregrinus, *Dictionary of Scientific Biography.*

14. A fascinating analysis of how people struggled to understand the phenomenon of the magnetic pole deviating from the geographic poles, and the fact that the magnetic pole seemed to vary in direction with time, has been given by science historian Stephen Pumphrey in an article entitled " 'O tempora, O magnes!' A Sociological Analysis of the Discovery of Secular Magnetic Variation in 1634." *British Journal for the History of Astronomy* 22 (1989): 181. It took many decades before anyone made sense of the fact that the two directions did not agree and to confirm that the magnetic pole direction really varied.

15. Samuel Eliot Morrison, *Admiral of the Ocean Sea.* (Boston: Little, Brown, 1942). The full story is given in this fascinating book.

16. Robert Norman, *The New Attractive.* (London: 1581; Reprinted Norwood, NJ: Walter J. Johnson, 1974).

17. Pumphrey, *"O tempora, O magnes!"*

18. Robert Norman, *The New Attractive.*

19. Ibid.

n 2 n
Clearing the Decks

A lodestone is a wonderful thing in very many experiments, and like a living thing. And one of its remarkable virtues is that which the ancients considered to be a living soul in the sky, in the globes and in the stars, in the sun and in the moon.

William Gilbert, De Magnete

IN 1600 the London physician William Gilbert, whose hobby was the study of lodestone, set the scene for understanding magnetism (Fig. 2–1). He was the first to confront the multitude of superstitions that surrounded this phenomenon and performed several experiments that revealed some of the properties of magnets. His magnificent treatise was entitled "On the Magnet: Magnetic Bodies Also, and On the Great Magnet the Earth; a New Physiology, Demonstrated by Many Arguments and Experiments." It is widely regarded as the first true work of modern science. In it he reported his greatest insight: that the earth was itself magnetic.

Primarily, Gilbert undertook the work of sorting out the wheat of fact from the chaff of fiction about magnetism. He suggested new perspectives that might be useful if one were to arrive at a satisfactory explanation for the phenomenon. In so doing he reminded his readers that

At an early period, while philosophy lay as yet rude and uncultivated in the mists of error and ignorance, few were the virtues and properties of things that were known and clearly perceived:

Figure 2–1. William Gilbert. Courtesy Burndy Library, Norwalk, Connecticut.

there was a bristling forest of plants and herbs, things metallick were hidden, and the knowledge of the stones was unheeded.[1]

In the course of man's early endeavors to explore natural phenomena, comments Gilbert, he chanced upon the discovery of the lodestone. "This, on being handled by metal folk, quickly displayed that powerful and strong attraction for iron, a virtue not latent and obscure, but easily proved by all, and highly praised and commended."

Thus the lodestone, after it emerged "out of the darkness and deep dungeons," became a respectable subject for discourse. Many famous philosophers had already tried to incorporate the phenomenon in their world views. Gilbert listed Plato "who thought the virtue divine" and Aristotle as having considered its virtue to attract iron and who had hinted at other properties as yet "all undiscovered." These thinkers had contributed to a broad tally of misconceptions about the nature of magnetism that determined how people before Gilbert's time thought about the phenomenon. Thus it was believed that if a lodestone "be anointed with garlic, or if a diamond was near, it does not attract iron." Gilbert tested these notions *by experiment* and proved them to be false.

Lodestone had also been considered as useful to thieves, so wrote Gilbert, as a love potion, a cure for gout and spasms, and "that it makes one acceptable and in favor with princes." It could remove sorcery from women and put demons to flight. It had the power to reconcile married couples. As with so many of human superstitions, those concerning lodestone were supposed to satisfy our deepest needs and wildest fantasies. Thus we can imagine earnest peasants dangling lodestones on ropes into deep wells after they pickled the rock in the salt of a sucking fish, because that was supposed to imbue the magic stone with the power to attract gold.

As so often in his book, Gilbert presages the titillation of modern tabloids. "With such idle tales and trumpery do plebian philosophers delight themselves and satiate readers greedy for hidden things." The readers of such tales were accused of being "unlearned devourers of absurdities." How little things have changed! Today, in the realm of astrology or the belief in UFOs as extraterrestrial spacecraft, we still find this continual need and a persistent craving for absurdities to delight and "satiate readers greedy for hidden things."

It is for revelation of previously hidden things that the mind strives when it expresses curiosity. And so we find that the first answers proposed to life's great mysteries lie in the realm of superstition for the simple reason that primitive minds could do no more than invent answers to questions that would remain enigmatic for several thousand more years. What other option was there? Ask why the sun moves across the heavens during the day and it is only natural to invoke a god riding a chariot. In ancient times no one knew what else the sun could possibly be, least of all just another star about which the earth was in perpetual orbit. Ask why lightning flashed and an answer couched in terms of the known (a god, or a searchlight perhaps) seemed to suffice, for a while at least. In human history, myth and superstition must have been used for countless millennia to account for natural phenomena.

Gilbert, without appreciating the great step he was taking, set the scene for a new world, one in which superstition would be transcended forever, or at least in the realm of inquiry that concerned the physical universe. First he systematically cleared the decks of ignorance and provided the base for understanding the nature of the phenomenon. He culled wild beliefs about the suspected power of lodestone from ancient writings. For example, taken with sweetened waters, "three scruples weight," lodestone was supposed to expel "gross humors." Gilbert didn't agree. "Others relate that lodestone perturbs the mind and makes folk melancholic, and mostly kills."[2] Some did not think it deleterious to health and even thought that it might be the elixir of life. Apparently lodestone could do almost anything, but that was, according to Gilbert, only because the ancients were ignorant of the true causes of things. "The application of a lodestone for all sorts of headaches no more cures them (as some make out) than would an iron helmet or a steel cap."

Iron itself was also regarded as medicinal. "It is given chiefly in cases of laxity and over-humidity of the liver, in enlargement of the spleen, after due evacuations; for which reason it restores young girls when pallid, sickly, and lacking color, to health and beauty. . . ." Apparently iron was often used, well before 1600, to combat anemia, perhaps the only one of the host of claims regarding the power of iron or lodestone that survives today.

Despite the efforts of many well-meaning men, Gilbert said that they were unsuccessful in finding an explanation for the lodestone because they were not "practised in the subjects of nature, and being misled by certain false physical systems, they adopted as theirs, from books only, without magnetical experiment, certain inferences based on vain opinions, and many things that are not, dreaming old wive's tales."

Although the whys and whats of magnetism were unknown, compasses had been in use since the thirteenth century, in particular in China. The reason that the compass needle pointed due north—or close to geographical north as determined by the motion of the stars—was believed by some, including Roger Bacon, to indicate a source of attraction in the heavenly vault itself. "Or that there is a magnet-stone situated under the tail of the Greater Bear," as one astrologer stated.

However, the end of superstitions about the lodestone lay at hand. Gilbert may have been intuitively aware of this when, in terms of poetic grandeur, he wrote:

> But the magnetick nature shall have been disclosed by the discourse that is to follow, and perfected by our labors and experiments, then will the hidden and abstruse causes of so great an effect stand out, sure, proven, displayed and demonstrated; and at the same time all darkness will disappear, and all error will be torn up by the roots and will lie unheeded; and the foundations of the grand magnetick philosophy which have been laid will appear anew, so that high intellects may be no further mocked by ideal opinions.[3]

He wondered why nature had been so stingy as to provide only a small number of metals and attacked astrologers for referring the individual metals to specific planets. Gilbert was also way ahead of his time in seeing through the mumbo-jumbo of astrology. He said that one of that ilk, Lucas Gauricus, covered many "shameful pieces of folly with a veil of mathematics." In general he considered astrologers "simple-minded and raving" and appealed to thinking individuals to set aside books filled with ignorance and to seek answers for themselves, a laudable admonishment that has echoed all but unheard through the ages since.

> Deplorable is man's ignorance in natural science, and modern philosophers, like those who dream in darkness, need to be aroused, and taught the uses of things and how to deal with them, and to be induced to leave the learning sought at leisure from books alone, and that [which] is supported only by unrealities of arguments and by conjectures.[4]

He railed against the philosophers of previous ages, none of whom ever undertook what he did, the direct study of magnets.

> Every good and perfect piece of iron, if drawn out in length, points North and South, just as the lodestone or iron rubbed with a magnetical body does; a thing that our famous philosophers have little understood, who have sweated in vain to set forth the magnetic virtues and causes of the friendship of iron for the stone.[5]

Like Peregrinus before him, he found that a piece of wrought iron had a north and south pole (Boreal and Austral) and discovered that when cut in half each piece had two poles as well. He studied the physical motions involving magnets; *direction* toward the poles of the earth, which would include attraction and repulsion; *deviation* or *variation* of the magnet's orientation from true north; *inclination* to the vertical observed when a magnet was allowed to swing in the vertical plane; and finally *revolution*, the rotation of the earth with respect to the stars.

Inclination showed that the force of attraction was rooted within the earth, which helped convince Gilbert that the planet was a giant lodestone, consistent with there being so much iron beneath the ground. He considered magnetic attraction, or at least what others labeled as such, but preferred to use the term "coition." According to Gilbert: "The word attraction unfortunately crept into magnetic philosophy from the ignorance of the ancients; for there seems to be a force applied where there is attraction and an imperious violence dominates." Coition, however, implied a running together of a gentler sort.

The key question was what caused coition. Therein lay the essence of the study of magnets for the previous two thousand years. Clearly lodestone had a near magical ability to attract iron over a distance. But why? The idea of vapors issuing from the lodestone had long been in fashion, although their role in the act of coition was necessarily vague. An attempt of Johannes Costaeus of Lodi to explain the phenomenon was quoted in detail by Gilbert:

> There is mutual work and mutual result, and therefore the motion is partly due to the attraction of lodestone and partly to a

spontaneous movement on the part of the iron: For as we say
that vapors issuing from the lodestone hasten by their own nature to attract the iron, so also the air repelled by the vapors,
whilst seeking a place for itself, is turned back, and when turned
back, it impels the iron, lifts it up, as it were, and carries it along;
the iron being of itself also excited somehow.[6]

Gilbert criticized this explanation as not being in accord with
the facts, in particular with regard to the issuance of vapors and
turning them back. Instead he imbued the earth with a magnetic soul, a continued adherence to a superstition, this one of
his own making. He also dismissed rumors that claimed that
iron was attracted because it was cold. That was considered to
be "a chilly story, and worse than an old wive's tale." Short shrift
was also given to claims that the lodestone was alive and that
iron was its food. All in all, Gilbert was not impressed with any
theories the ancients had conjured up to account for magnetism. He felt it best that the old ideas be relinquished "to the
moths and the worms."

Instead, he formulated his own explanations and made the
point that in seeking an explanation it was not just magnetism
that was in need of an answer. When amber was rubbed it had
the ability to pick up a variety of substances such as pieces of
cloth. This was called *electricks* and is due to what we now call
static electricity.

Gilbert suspected that in the case of magnets some aspect of
inherent and primary form was involved. That was unlike amber, whose powers of attraction could be manifested only after
doing something to it (rubbing it vigorously). It was but a small
step to conclude that given the existence of a primary form of
attraction in the lodestone there had to be something similar in
each globe such as the earth, sun, moon, and the stars.

Wherefore there is a magnetic nature peculiar to the earth and
implanted in all its truer parts in a primary and astonishing manner; this is neither derived nor produced from the whole heaven
by sympathy or influence or more occult qualities, nor for any
particular star; for there is in the earth a magnetic vigor of its
own, and a small portion of the moon settles itself in moon-man-
ner toward its termini and form; and a piece of the sun to the

sun, just as the lodestone to the earth and to a second lodestone by inclining itself and alluring in accordance with its nature.[7]

So Gilbert pictured magnetism to be inherent in the magnetized object, and the result of its form. He was not blessed with the technical means of exploring the nature of that form, however. At best that capability lay many centuries in the future. One thing he did realize was that if something was given off by magnets it had to be vaporous, diaphanous as it were, in order to enter into iron.

He confessed that the notion of the lodestone having a soul, attributed to Thales of Miletus, was not so far off the mark. Just as a soul was believed to reside in a body, so something resided within the magnet that could be made to depart under appropriate conditions. This made the phenomenon even more fascinating. This inherent form (soul) lay dormant in iron and was thought to be stimulated into existence by the mere proximity of the lodestone.

Gilbert began to experiment systematically to discover what lodestone could or could not do, and what happened when a piece of iron was magnetized. Lodestone attracted a piece of iron with one end and repelled it from the other. A piece of iron magnetized by the lodestone would lose its power when heated to incandescence. He discovered that beating wrought iron with a hammer induced magnetism (Fig. 2–2). (Today we know that such impacts cause the molecules to be jarred and for an instant they respond to the earth's magnetism, which pulls at them to produce a net alignment, and hence a residual magnetization in the iron.) He correctly inferred from the fact that heating caused magnetism to disappear that "fire destroys the magnetic virtues in a stone, not because it takes away any parts specially attractive, but because the consuming force of the flame mars by the demolition of the material the form of the whole; as in the human body the primary faculties of the soul are not burnt, but the charred body remains without faculties." This analogy left a lot to be desired, but the notion of the destruction of the interior form was essentially correct.

As he carried out experiments to find what caused magnetism he collected a great deal of data, which, when taken together, might allow him to understand the phenomenon. He

Figure 2–2. An illustration from Gilbert's book, De Magnete, *showing how hot iron, when beaten, can be made magnetic. Courtesy Burndy Library, Norwalk, Connecticut.*

was on the right track but unable to move very far along it. The properties of atoms and molecules were not yet known. The concept of an atom, essential for the understanding of magnets, would not be developed for another three hundred years. Gilbert, like Peregrinus before him, was way ahead of his time, at least as far as his ambitions toward understanding magnetism were concerned.

We can look back on Gilbert's experimentation with the benefit of hindsight and realize that there was one leap of intuition that he did not make, one that was blocked because he treated the lodestone as the primary object. Lodestone was a lump of magnetite, a form of iron oxide, gathered together in a stone that happened to manifest residual magnetism. The more perfect magnet is an elongated piece of iron that has been magnetized by striking it or by stroking it with a lodestone, the imperfect magnet. It was not the last time a scientist would spend his or her time pursuing a secondary phenomenon because the primary one lay hidden. Yet, by concentrating his attention on a lodestone of spherical shape, Gilbert did intuit the connection with a magnetized earth, and he appreciated the fact that near the terrestrial poles compasses no longer functioned as direction indicators because there they would point straight down to the ground.

The spherical lodestone he used in his experiments was called a terrella, which sent "out in an orbe its powers in proportion to its vigor and quality." An orb of virtue was defined as a volume of space around the terrella or magnet in which the presence of magnetism could be sensed. (Today we call this the "field" around the magnet, and refer to the magnetic field as defining this sphere of influence. See Chapter 7.)

Clearly Gilbert was a great scientist, one of the first, and he systematically studied the phenomenon that so fascinated him. He observed that a piece of iron that had been magnetized and then demagnetized through heating would slowly regain some magnetism, "having acquired some power from the earth." This was a remarkable conclusion, as was his observation that unlike heat, which is conducted slowly from one end of a heated rod to another, magnetism instantly occurred at the far end of a rod being magnetized at the near end. These were valid observations, but the explanations for those manifestations required the invention of measuring instruments, without which the scientific study of magnetism could not progress.

Gilbert's greatest contribution may have been that he so clearly separated superstition from facts about magnetism and thereby cleared the decks of a great deal of confusion. Other scientists, unhampered by what Gilbert called old wives' tales, could now move forward.

Significant progress, however, lay well in the future. In the meantime, others were drawn into the study of magnetism, not all of them allies of Gilbert. For example, in 1629 Niccolo Cabeo published a book called *The Magnetic Philosophy* in which he tried to prove Gilbert wrong. His said that magnetism was akin to "an electric," which, when rubbed, created heat that forced air away from it. Then, in the low pressure area so generated, things would be drawn together. This was argued until 1675, when Robert Boyle developed a vacuum system within which magnets still functioned. Gilbert's conception of the earth as a spinning magnet also met widespread disapproval, because most people did not believe the earth rotated (see Chapter 3.)

More appreciative of Gilbert was William Barlow in his book, *Magneticall Advertisements,* published in 1616, subtitled "Divers pertinent observations, and approved experiments concerning the nature and properties of the loadstone: Very pleasant for knowledge, and most needful for practise, of travelling, or framing of instruments for travellers both by sea and land."[8] Barlow waxed lyrical in his praises of the magical rocks, which he said were justly admired because they expressed the "infinite power, and goodness of our God, who hath created so precious a jewel for the profitable use of man, and for the enlarging, and setting forth of his own glory."[9]

Barlow was an expert on how to magnetize needles, and he studied what happened when such needles were joined or cut. His book was important enough to deserve an addendum from Gilbert, who praised the author for "finding divers secrets" concerning the stone. The real secrets, however, would lie hidden until someone invented a means for making measurements of forces between magnets.

NOTES

1. William Gilbert, *De Magnete,* (London: Chiswick Press, 1900) p. 1.

2. Ibid., p. 32.

3. Ibid., p. 7.

4. Ibid., p. 28.

5. Ibid., p. 30.

6. Ibid., p. 62.

7. Ibid., p. 65.

8. William Barlow, *Magneticall Advertisements.* (1616; Republished by Da Capo Press, Amsterdam and New York, 1968).

9. Ibid.

∩ 3 ∩
On the Magnetical Philosophy

> The magnetic force is animate, or imitates a soul; in many respects it surpasses the human soul while it is united to an organic body.
>
> William Gilbert, *De Magnete*

FACTS are necessary to transcend points of view rooted in pure thought. Facts, however, are hard to come by. It was one thing for Gilbert to alter his point of view about magnetism; it was quite another for the study of the phenomenon to be placed on a quantitative footing. But since his book caused such a stir, readers wanted to make use of the ideas. This they proceeded to do with a vengeance. For at least half a century following the publication of *De Magnete* (in 1600), Gilbert's insights were at the heart of what came to be called the Magnetical Philosophy, which allowed a wide range of phenomena to be erroneously accounted for in terms of magnetism. For example, proselytizers of both the earth-centered (geocentric) and the sun-centered (heliocentric) views of the heavens were to draw upon it for inspiration.

When a particular phenomenon is understood for the first time, the thrill of the insight moves us to use the explanation as a panacea for all ills. Leaping up and shouting "Eureka, I have found it," enthusiastic converts apply their new understanding with little or no discrimination. It has been going on for thousands of years. Apparently we are easily enamored of our own ideas.

It is no different in science, although the scientist is sup-

31

posed to use his or her skills in a struggle to transcend first impressions, in particular to make sure that the insight is valid— that is, pertains to objective reality—rather than a purely subjective experience. First impressions have a way of misleading. One need only look at how various thinkers have confronted the motion of the heavens, of the stars in their course around the earth, a fact that was long known and misunderstood. Two millennia ago it was believed that the planets, sun, moon, and stars all moved around the earth in circles. That was believed because Plato had once said that circles were perfect, and obviously heavenly motion had to be perfect. The problem was that in order to describe planetary motion a single circle per planet did not work. But if you added a small circle to a larger circle (to produce an epicycle) things worked a little better. But with each passing century, astronomers found that they needed to add a circle to the circle upon the circle, and so on, until after about seventeen hundred years, models to account for heavenly motion were a confused mess of epicycles.

In Gilbert's time the epicyclic model for the solar system was beginning to collapse. Nicolaus Copernicus triggered the revolution with the publication of his book on the motion of the planets, *De Revolutionibus,* in 1543. He wrote that the sun was at the center with the planets orbiting it in circles (plus a few epicycles thrown in for good measure). The basic idea was itself not new. Back in about 250 B.C., Aristarchus of Samos suggested that it was not the stars that were in motion around the earth, but the earth that rotated beneath them. This gave rise to the illusion of stars going around the earth. But the earth had been kept at the center of the heavens for another seventeen hundred years for several reasons. Above all, the theory seemed to fit the observations, which were relatively crude, of course, since the telescope would not be invented until 1609. But it was also true that religious beliefs about the heavens gave humankind a central role. That implied that the earth, too, was central. So when Copernicus's world view came along, it was not a threat to anyone, because it was only a theory with no demonstrable facts to back it up. No one could prove beyond reasonable doubt that Copernicus was correct; at least not without much better observational data than were available at the time.

At this point Gilbert entered the picture, and to him it was obvious that Copernicus was right. In his book Gilbert added his own brand of logic to support the Copernican point of view that the earth was not at the center of the heavens. He did not mince words when he argued the case.

Ancient philosophers had pointed out that since the distant stars were moving around the earth, it implied the existence of a prime mover, *primum mobile,* which had to be, in Gilbert's skeptical words, "a universal force, an unending despotism, in the governance of the stars, and a hateful tyranny" because it controlled everything. "Surely that is a superstition, a philosophic fable, now believed only by simpletons and the unlearned; it is beneath derision; and yet in times past it was supported by calculation and comparison of movements, and was generally accepted by mathematicians, while the important rabble of philophasters egged them on."[1] Clearly he was not impressed by the notion. In fact, he was enraged that anyone could ever had been so stupid as to imagine the heavens to be rotating around the earth. Even if the primum mobile existed, what "mad force" lay beyond it? Surely the agent that determined the motion should abide "in the bodies themselves, not in space, nor in the interspaces." But to search for its cause was to lose track of the main issue, which to Gilbert was glaringly obvious. The earth had to be rotating. "This seems to some philosophers wonderful and incredible," he wrote scathingly. But was that any more incredible than to imagine the rest of the universe moving around the earth? Surely it was easier to picture that a small piece of it, our planet, was in motion. Gilbert said that if the reader could not accept this, he or she should bear in mind that it was "worse than insane" to imagine that an even larger mass, that of the primum mobile, was instead able to move around the earth. It was no use trying to wriggle out of the dilemma by imagining that the distant heavenly spheres weighed almost nothing, because if they were ethereal they would have no substance and would quickly shatter owing to their headlong speed.

To Gilbert the conclusions were obvious: "Let the theologues reject and erase these old wives' stories of a so rapid revolution of the heavens which they have borrowed from certain shallow philosophers."[2]

But he was not done yet. With grand and glorious cynicism, he pictured what it was that the philosophers and theologues (whom he hated so much) would have him believe. While everything from the largest to the smallest scale swings wildly around our planet, the earth was not "stayed in its place by any chains, by no heaviness of its own, by no contiguity of a denser or a more stable body, by no weights. The substance of the terrestrial globe withstands and resists universal nature."[3] He hinted that any reasonable person who gave this any thought— and Gilbert was a man of reason—would conclude as he had, that the earth moved on its own axis and that was that. One can almost picture how thunderstruck he felt knowing that for thousands of years no one else seemed to have realized the elegance of this argument.

The essence of the problem lay in the way the stars had to be moving if they went around the earth. As Gilbert put it, "It is then an ancient opinion, handed down from the olden time, but now developed by great thinkers, that the whole earth makes a diurnal [daily] rotation in the space of twenty-four hours."[4] If the stars did move around the earth in one day, they had to indulge in headlong motion to get all the way around. Compared with the planets, the stars were far away. If they were whirling so wildly, how could they possibly keep their relative positions? Also, why would they take heed of earth? These were good questions to which he offered his good answers.

"Astronomers have observed 1022 stars," Gilbert said, and hinted that there were many others not yet as closely studied. Where were they suspended? The stars were obviously very far away, "beyond the reach of eye, or man's devices, or man's thoughts." Each star had to be fixed and the center of its own system, which, if it had motion, would be about some local center. There was no difference in the properties of distant stars; their color and brightness were roughly the same whether they appeared high above the equator or near the celestial poles. This meant that the stars were similar enough to be considered a large family. And therein lay the problem. How could that large and very distant family pay any attention to the existence of the earth? It would not do so. This argument led Gilbert to draw the only reasonable conclusion: the earth was rotating on its own axis and that gave rise to the illusion that the stars move around the planet.

The London physician was not one to hold back once he had seen the light. "And now, though philosophers of the vulgar sort imagine, with an absurdity unspeakable, that the whole heavens and the world's vast magnitude are in rotation," he wrote. But that didn't explain why the earth rotated. In the search for an answer, he gave thanks to the Creator for having set it up this way; hardly a scientific answer. But because the earth had two poles, like the lodestone, he could imagine that it had a natural axis in space, a north–south alignment—verticity, he called it. Then all you had to imagine was that some cosmic magnetic force pulled on the earth to cause it to spin around the axis.

To Gilbert the connection between magnetism and the heavens was then easy to make. Given that the earth was magnetic and that magnetic effects reached their tentacles into space surrounding the magnetized body, a connection between the earth's magnetism and that of the heavens had to exist. This theory rested on his speculations that the lodestone is magnetically aligned with the earth's own magnetism, because the lodestone had lain for so long in the bowels of the earth and hence acquired its magnetization from the planet.[5] By analogy, the earth must have acquired its magnetization from space.

His almost mystical explanation for the earth's rotation revealed that his private cosmology still rested firmly in the laps of the gods. According to science historian J. A. Bennett, "What Gilbert meant by 'magnetism' was the expression of spiritual influences whose animating presence was felt throughout Nature."[6] This caused Gilbert to describe the earth's magnetic properties in glowing terms:

> By the wonderful wisdom of the Creator, therefore, forces were implanted in the earth, forces primarily animate, to the end the globe might, with steadfastness, take direction, and that the poles might be opposite, so that on them, as at the extremities of an axis, the movement of diurnal rotation might be performed. Now the steadfastness of the poles is controlled by the primary soul.[7]

Based on his experiments with the terella, Gilbert thus reasoned that if the earth could be moved so that its axis pointed away from north, it would inevitably swing back to its correct alignment. The mystical magnetic influences that he had done so much to rescue from the maw of superstition would then cause the earth to point again toward Cynosura, the key direction, the center of attraction associated with Ursa Minor, specifically the pole star, Polaris. In other words, just as the lodestone floating in water would align north–south, the earth, floating in air as it were, would also swing to align north–south. This notion, that the earth's magnetism was linked to the heavens, set a bandwagon rolling down the avenue of human thought for a half century.

Gilbert's misguided insights played an important role in the intellectual climate of the time. Science historian Martha Baldwin described Gilbert's contribution: "By adding magnetic motions and magnetic souls to the forces impelling and ordering the heavens, Gilbert significantly enlarged the field of astronomical enquiry and debate in the seventeenth century."[8]

Then Galileo Galilei appeared on the scene and turned his telescope to the heavens. In 1609 he discovered the moons of Jupiter, which showed that Jupiter was a local center of attraction for its moons. He saw that Venus displayed phases, which meant that it had to be orbiting the sun inside the earth's orbit. His observations proved that the earth was *not* at the center of *all* motion. At the same time, Johannes Kepler was discovering the laws that determined how planets moved around the sun, the essence of which was that the planets moved around the sun in elliptical orbits. The work of these two astronomers suddenly placed Copernicus's ideas on a firm scientific footing and profoundly shocked those in the civilized world who cared about such things. Reacting particularly strongly was Rome, given that the Church was rooted in ancient dogma, which, in turn, attached great significance to man's (and hence earth's) central position in the universe.

Even as the Church battled to suppress the news—by muzzling Galileo with house arrest and ordering that he desist from spreading his heresies—inquiring minds began to wonder what actually kept the planets in their orbits. It was one thing to recognize that the sun was at the center of the solar system; quite another to explain how it was possible for the planets to keep traveling around and around without either falling into the sun or being lost in space. What held them in its embrace? Thanks to Gilbert's work, the answer was ready at hand. Magnetism had to be responsible.

Gilbert had long appreciated how important it was for the sun to wander up and down relative to the horizon during the year to give rise to the seasons. Without this motion,

> the sun would ever hang with its constant light over a given part, and, by long tarrying there, would scorch the earth, reduce it to powder, and dissipate its substance, and the uppermost surface of the earth would receive grievous hurt; nothing of good would spring from earth, there would be no vegetation; it could not give rise to animate creation, and man would perish.[9]

How was disaster to be avoided? By making use of wondrous magnetical energy of course. It would allow the earth to seek out the sun, again and again, and prevent it from pointing toward where there was no day and night.

Gilbert explained it this way:

> The sun (chief inciter of action in nature), as he causes the planets to advance on their courses, so, too, doth bring about the revolution of the globe [the earth] by sending forth the energies of his spheres.[10]

The sun was supposed to pull the earth around in the way a magnet that was moved around a terella suspended in water caused it to rotate. Clearly something had to be *pulling* the earth around and that something had to be the same force that caused a compass needle to point north. It had to be the magnetic pull of the sun that caused the earth to rotate.

Gilbert's efforts to account for the rotation of the earth represent one of the first attempts to explain this motion, but the details of his theory could have been covered by the same umbrella of scorn he heaped upon others. He was up to his neck in a new set of wild beliefs bordering on superstitions. But who could blame him for enthusiastically sounding the drum to proclaim the usefulness of his magnetical philosophy?

He even tried to explain, in tortured fashion, why a day is exactly twenty-four hours long. This was supposed to depend on the magnetic interaction between the earth, moon, sun, and stars, all known to be at different distances and hence exerting subtly different pulls on one another. Gilbert did admit that he was hardly in a position to comprehend the nature of this force;

it was enough that he had found the earth to be magnetic. But once that was recognized, everything else followed, although it was not obvious why. And thus was born the magnetic philosophy, which held center stage in astronomical debate for the next fifty years.

With hindsight we recognize that most if not all of Gilbert's astronomical arguments were wrong. Yet we can also appreciate how enormously difficult it was for anyone to comprehend the fundamental motion of the heavens. These would only be understood by successive generations, including our own, but we tend to forget that such phenomena were once very mysterious indeed and that they posed tremendously difficult intellectual challenges for those who thought about them.

A surprising consequence of Gilbert's efforts to account for the earth's motion using the magnetic pull of the sun was to make fashionable the idea of the existence of forces capable of acting over a distance. This paved the way for future acceptance of a universal theory of gravity, which involved a force not unlike magnetism. For a period during the seventeenth century, the terms gravitational and magnetical were often used interchangeably.[11] Bennett has said, "The magnetical philosophy in England has not been given the attention it deserves— deserves not only in its own right, but also for the crucial role it played in the history of cosmology. It is part of the explanatory background to the emergence of Newtonian theory in the 1680's." Although incorrect in its details, Gilbert's suggestion of "the concept of attraction was too useful to be given up lightly, an instinct which in time proved to be sound."[12]

Gilbert's magnetical philosophy provided the setting for Newton to propose the concept of gravity, and Galileo and Kepler both made use of magnetism to account for heavenly motion. For a while it was fashionable to use magnetism to bolster virtually every cosmological argument, no matter which side of the geocentric/heliocentric debate you were on. The magnetical philosophy was such an all-purpose structure that it was used by both sides to shoot down the arguments proposed by the other. Magnetism was used to prove that the sun was at the center of the heavens. It was also used to prove that the earth was at the center.

Kepler and Galileo found the magnetic concepts of great

use in their own struggles to understand what kept the planets in orbit. Kepler said so explicitly when he wrote that he had "placed a celestial rooftop upon the magnetical philosophy of Gilbert, who himself has built the terrestrial foundation."[13] It all fitted so well that members of the opposition, many of them clerics struggling to reconcile Galileo's heretical claims with the teachings of the Bible, were forced to take a different approach to rescue their dogma from extinction.

One who took up the challenge was a Jesuit, Niccolo Cabeo (1586–1650), who began by pointing out that magnetism acted over only a short distance. Once you accepted that as a fact everything else followed. To him it meant that "gravitational forces held the earth in the middle of the universe so that it could receive essential and vital influxes from the stars as they coursed around it. If some alien force or impetus produced by the swiftly rotating heavens should dislodge earth from its central position, the magnetic force would quickly restore the earth to its original position in a proportionate manner."[14] So gravity held earth in place at the glorious center of everything, and magnetism made sure the earth continued to maintain the orientation of its axis. It was all very elegant and very wrong.

Others joined the fray. A creative description of why the earth was magnetic was proposed by a French Jesuit, Jacques Grandami (1588–1672):

> Although gravity causes the earth to stand in the center of the world, it is not able to impede its circular motion around the center, especially against the daily agitation of all the sea waters in the changing tides and in violent storm. Thus it is that another quality is added and assigned to immobility. . . . This quality is sufficient for effecting this immobility and for restoring the earth's situation with the poles of the sky if by chance it should be disturbed. I call this quality the magnetic quality since in magnetic bodies the rest and constant immobility on the meridian line (or near it) are seen everywhere.[15]

Bennett has pointed out that "Grandami claimed his magnetic proof of the earth's immobility was rooted in physical and moral truth."[16] Yet the moral of our story is that when one uses beliefs or cherished dogma to account for nature's wonders, one is inevitably led up the garden path and left with nothing, not a single idea that can stand the test of time, if for no other reason than that the original ideas that gave rise to the dogma were fictions of human imagination to begin with. They were fictions because they were not founded on observations of nature under experimental conditions. It was here that Gilbert's work stood out so massively, even if his *interpretations* of his observations often left much to be desired. He made observations that revealed certain qualitative patterns in regard to the nature of the forces exerted by lodestone. Little did he know that it would be a century or more before the necessary quantitative aspect would be added to observations, a factor absolutely necessary before anyone could begin to probe beyond nature's appearances.

In the meantime another Jesuit, Althanasius Kircher (1602–1680), took up the challenge of using the magnetical philosophy to shoot down the misguided notions of heliocentrists like Galileo and Kepler.[17] He showed that the magnetical theory could not work, because if the earth were magnetic and the moon were made of iron it would be pulled toward the earth, which it wasn't. Mountains containing iron should slide inexorably deeper into the earth, which they didn't. Men working with iron tools would find it impossible to lift them, and obviously blacksmiths were doing fine. Thus the whole notion that the earth was magnetic had to be wrong. But several fundamental and incorrect assumptions were hidden in Kircher's arguments. Mountains are not made of iron. Even if they were, the earth's magnetism would be too weak to cause such a mountain to move. (The same may not be true on a neutron star, where the magnetic field is a trillion times as strong as on earth and mountains of iron would be dragged about at the whim of the magnetic force.) And the earth's field is only strong enough to swing a delicate compass needle. It does little to orient iron tools let alone cling to them with such force as to make them impossible to lift.

Baldwin commented that "Kircher's motive in challenging Gilbert and Kepler went beyond an impartial interest in magnetic science; with religious zeal he considered their scientific fallacies 'pernicious to the Christian Republic and dangerous to the faith of the Church.' Gilbert and Kepler, like other Copernicans, had failed to take Biblical astronomy seriously."[18]

The seventeenth century marked a move away from reliance on superstitions, outdated beliefs, or biblical myths for natural phenomena. The transition is not yet complete. That is why fundamentalists still struggle with the facts of evolution. It is very difficult for the human mind to accept a new truth, especially one that transcends a long-held belief.

What happened back in 1600 or so was that human beings began to seek answers to their questions by confronting nature directly. Above all, they paid attention to the truths (facts) revealed in experimental situations. This approach was completely at odds with the ancient technique, finely honed over thousands of years, that relied on verbal argument to prove a point without heeding the messages found in natural phenomena, and certainly not by relying on experiment. In the specific case we have just considered, Martha Baldwin put it this way, "Kircher saw in the novelties of Gilbert's and Kepler's magnetic astronomies an intolerable temerity of the human mind to limit divine omnipotence and to claim to know fully the ways of God."[19] But that is precisely what has been going on every since. Scientific questioning has led to discoveries that set limits in areas where ardent believers used to picture God as capable of anything. If there is one thing the scientific quest has turned up it is that there are natural laws that determine what is possible and what is not possible in our physical universe. This would suggest that if God (should any such entity exist) created the laws, that God would surely be as subject to the laws as anything else in space-time.

Despite the shortcomings of the magnetical philosophy, Gilbert was a pioneer in showing a new way to proceed to uncover nature's deepest secrets. The road ahead now lay wide open. The quest would prove to be a long one. Even if Gilbert had overestimated the usefulness of his interpretations, his insights revealed the presence of order in what had for so long appeared to be a chaotic world. If nothing else, that is what science is about: recognizing order in the world of phenomena and extracting from that order expressions of underlying laws that account for the way things are.

NOTES

1. William Gilbert, *De Magnete,* Part VI, Chapter III. This and the following references to Gilbert's work refer to a relatively recent translation by P. Fleury Mottelay. (New York: Dover Publications, 1958) p. 322.

2. Ibid., p. 325.

3. Ibid.

4. Ibid., p. 316.

5. See Part VI of *De Magnete* for details of Gilbert's astronomical thinking.

6. J. A. Bennett. "Cosmology and the Magnetical Philosophy, 1640–1680." *Journal of the History of Astronomy* 12 (1980): 165.

7. Gilbert, *De Magnete,* p. 328.

8. M. R. Baldwin, "Magnetism and the Anti-Copernican Polemic." *Journal of the History of Astronomy* 16 (1985): 155.

9. Gilbert, *De Magnete,* p. 333.

10. Ibid.

11. Bennett, "Cosmology," p. 174.

12. Ibid., p. 176.

13. Ibid., quoted on p. 156.

14. Ibid., p. 157.

15. Ibid., quoted on p. 167.

16. Ibid., p. 168.

17. Kircher invented the first magic lantern (slide projector).

18. Bennett, "Cosmology," p. 160.

19. Ibid.

n 4 n
Let the Experimentation Begin

A man who "understood a lot about the electricity of women."

> A description of Alessandro Volta,
> *Dictionary of Scientific Biography*

T HE first step on the long journey of subsequent discovery regarding the nature of both magnetism and electricity was taken in 1660, when Otto von Guericke (1602–1686) built the first electric generator that made plenty of sparks when you needed them. His device consisted of a large ball of sulfur mounted on a long shaft with a hand crank. When the sulfur ball was rotated at high speed and a cloth applied, sparks would leap about, particularly between a spark gap connected to two brushes that touched the spinning sphere. Von Guericke carried this device from place to place in his laboratory and applied the electricity where he wanted it.

Serious experimentation into the nature of virtually any physical phenomenon could begin only when two conditions were fulfilled. First, a lot of material had to be available on which to experiment, and, second, measuring devices were needed to give quantitative information about the phenomenon.

As far as magnetism was concerned, lodestones were plentiful and they could be used to magnetize needles of iron. However, the strength of such needles could never be greater than that of the lodestones themselves. As to electricity, it was all very well to rub amber and find that it picked up pieces of straw. But so what? How could one begin to experiment with either of these phenomena? With the benefit of hindsight, we know that what was needed was lots of electricity, more powerful magnets, and devices for measuring various properties of electricity and magnetism. Only then would experimentation begin in earnest.

An accidental discovery made by Peter van Musschenbrock (1692–1761) at the University of Leyden (now Leiden) in the Netherlands allowed serious experimentation with electricity to begin. One day he was playing with a glass globe friction machine (not unlike von Guericke's generator) when he fed the electricity along a wire through the neck of a glass jar. His assistant was holding the jar. After the sparking stopped, van Musschenbrock reached out and touched the wire and received a very unexpected shock. Then he sent electricity straight into the empty jar, which he held in one hand. After the generator stopped he touched the wire leading in through the neck. "Suddenly I received in my right hand a shock of such violence that my whole body was shaken as by a lighting stroke," he wrote. "The arm and body were affected in a manner more terrible than I can express. In word, I believed that I was done for."[1] Clearly, he wasn't. Instead he had found a way to pour electricity into a bottle and keep it stored for a while.

The Leyden jar acted as a capacitor—a device for storing electricity—and it allowed him to carry the stuff around the laboratory (although he still had to crank up the friction generator to fill the bottle).

As every experimenter knows, it is one thing to have available ample quantities of the stuff you wish to play with, but it is quite another to know *how much* of it you have. It is all very well to have electricity available in the wall sockets in your house, but to use it in quantitative experimental situations you need to know how much of it is passing into the apparatus. (Thanks to an appropriate measuring device, power companies can figure out how much each household must pay for electricity each month.)

Real progress toward understanding the nature of electricity and magnetism required that someone invent a measuring device that would give an indication of the force acting between charges or magnets. The first such device was the torsion balance invented by Charles Augustus Coulomb (1736–1806) in

France. He used it to measure the force between electrical charges and showed that the force dropped off inversely as the square of the distance separating them. This meant that when the distance between two charges was doubled, the force dropped to a quarter of its previous value (triple the distance and it drops to one-ninth, and so on).

Nearly two hundred years had elapsed since Gilbert began to set the record straight and freed the lodestone from the grip of superstition. Yet, very little progress in understanding magnetism had been made in that time. At best, people were left in a state of uncertainty, if they ever gave the question a thought, but that situation was about to change. Coulomb's invention was the first breakthrough. The next important step was largely due to frogs.

There is an idealized image of the typical scientist who logically proceeds to experiment, knowing ahead of time what he or she will find. This is a myth. Pioneers, whether migrating west across the plains of America, or exploring the frontiers of knowledge, seldom if ever know what lies ahead. This becomes dramatically obvious as we watch the next player on the scene, Luigi Galvani (1737–1798), professor of anatomy at the University of Bologna, stumble into his famous frog's leg caper. His accidental discoveries and subsequent experiments were the next major step on our journey of discovery.

In 1786 Galvani had been using frog legs as part of certain anatomical studies. On one side of his laboratory lay a pair of frog legs still attached to the vertebral column of the animal. On the other side stood a friction machine, an electrical generator similar to the device invented by von Guericke (Fig. 4–1). One day Galvani's assistant, Giovanni Aldini, happened to touch the frog legs with a scalpel and was stunned to see them convulse. Galvani repeated the experiment many times and found that the effect was most pronounced when the electricity generator was being cranked to produce sparks. When he touched the metal part of the scalpel to the frog's leg, electricity produced by the machine apparently triggered muscular activity. But nothing happened when Galvani held the scalpel by its bone handle. He also found that when a leg was placed inside a glass tube and held near the generator it contracted without any direct contact being made with it at all.

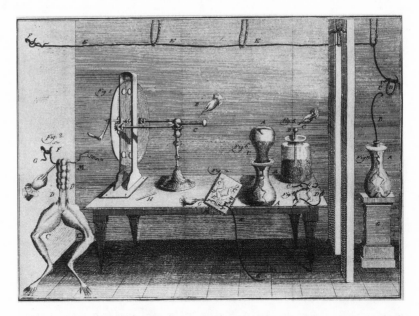

Figure 4–1. Galvani's laboratory with the electrostatic generator, which when turned triggered convulsions in the amputated frog's legs lying at the end of the table. The illustration is from A. Galvani's book Abhandlung über die Kräfte der thierischen Elektrizität, *1793. Courtesy Burndy Library, Norwalk, Connecticut.*

When one reads about Galvani's subsequent experimentation, one thing stands out clearly: with his scalpel he must have decimated the Italian frog population. He once attached frogs' legs to all the door hinges in his house and arranged a circuit so that when the electrostatic generator was activated all the legs leaped in unison. On another occasion he suspended frogs' legs from a large circular conductor surrounding the machine, and when it was cranked "he was rewarded by the sight of all the legs jumping together."[2]

That was nothing! He tried experiments on chickens and sheep and made their carcasses twitch. Rumor spread that if he could create such miracles perhaps he could bring back the dead, rumors no doubt fed by the antics of Aldini. He, together with Galvani and his wife, formed a trio that experimented in a laboratory at the Galvani home where Aldini found a way to produce muscular spasms in the severed heads of sheep, chickens, and oxen. He could make unhearing ears twitch and unseeing eyes blink. This was all too good to be true and Aldini took the show on the road. He traveled all over Europe, and in public made the carcass of an ox kick its feet—to the great consternation of the onlookers. It was not long before he turned his attention to human corpses, and one notable experiment involved the body of an executed murderer.

> After the body had lain for an hour exposed in the cold it was handed over to the President of the London College of Surgeons who co-operated with Aldini in making numerous observations to determine the effect of galvanism [the name given to electricity in those days] with a voltaic column[3] of one hundred and twenty copper and zinc couples.[4]

As in a horror movie, the corpse moved its arms and legs as though lifting weights or walking.

Meanwhile, back home, one of Galvani's great achievements involved his frogs'-legs detector of lightning in thunderstorms. A vertical rod of iron was mounted in the open air and insulated from the ground. To this he attached a set of frog legs connected to the rod. Their other ends were connected to a wire that ran down into a well. The results were marvelous. When lightning flashed, the legs twitched instantly, well before the thunder could be heard. Because artificially generated electricity and lightning produced the same effects, it meant that lightning was a form of electricity. This, in itself, was a major discovery.

Galvani fastened brass hooks into the spinal cords of prepared frogs and suspended the hook from an iron railing that surrounded a hanging garden at his house.[5] One day the legs twitched during a thunderstorm, even though the iron rod he had previously used to collect the invisible electrical emanations from the lightning was not in place. He later noticed the same twitching when the "sky was quiet and serene," but this effect was very infrequent. No matter what he tried, he couldn't find what controlled it. Imagine how many frogs' legs were used during his patient wait for the phenomenon to repeat itself. Finally he began to suspect he had been seeing things. "In experimenting it is easy to be deceived," he concluded, "and to think we have seen and detected things which we wish to see and detect."[6] So he gave up.

Now the scene was set for what must surely be one of the great oversights in the history of science. It would be straightened out by another Italian who turned it into an important invention that opened up the study of electricity and magnetism to widespread experimentation on a scale not previously imagined.

Galvani's major oversight began when he and a new assistant found they could make a dead frog dance without using the electric generator. Galvani held a frog's body by a hook and let its feet touch the top of a silver box. When he touched the box with a metal rod he held in his other hand, the corpse twitched. When his assistant used the rod to touch the box, nothing happened, at least not until he and Galvani held hands (Fig. 4–2). Then the electrical circuit was made complete and the dead frog danced.

At this point Galvani decided that the completion of the circuit allowed "animal electricity" to flow out of the frog's body. The muscles must act like small Leyden jars, he thought, storing electricity until it was allowed to discharge. This seemed to him to wrap things up nicely, but he was wrong. Had Galvani paid closer attention he might have made the giant leap that revolutionized the way electricity could be produced (in which case we might now have 110 galvanis of electricity in our homes, instead of 110 volts!). But that step was left to someone else.

Alessandro Volta (1745–1827) was the son of an Italian family that had risen to nobility (Fig. 4–3). According to Michael Faraday, Volta was outgoing and jovial, "a hale elderly man, very free in conversation."[7] Volta's passion for the study of electricity was well developed by the time he was eighteen years old, when he suggested that electrical phenomena such as that produced by rubbing silk by hand or with glass (what we now call static electricity) resulted from some attractive force that existed within the objects being rubbed. Under normal conditions there was supposed to be a balance between particles attracting one another within an object. When the object was rubbed, some of the particles were displaced, so that the balance was disturbed and a net attraction resulted. An electrical "fluid" was redistri-

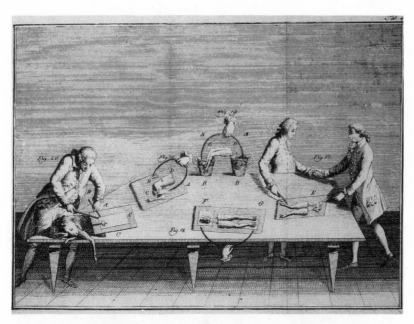

Figure 4–2. At the right, Aldini and Galvani are holding hands while touching rods of different metals to the amputated frog's legs to produce a convulsion. At other places in the sketch various experiments performed by Galvani and his assistant are illustrated. The illustration is from A. Galvani's book Abhandlung über die Kräfte der thierischen Elektrizität, *1793. Courtesy Burndy Library, Norwalk, Connecticut.*

buted in the object and it was the interaction between this fluid and common matter that gave rise to the phenomenon of electricity, or so Volta thought.

In 1791 he learned about Galvani's experiments with frogs' legs and that Galvani had suggested that when he linked the muscle in an electrical circuit some sort of "nerveo-electrical" fluid that accumulated in the muscles was released. Volta scoffed at the notion and thought it "unbelievable," although he did concede that what Galvani had found was "miraculous." At the same time he did little to hide his low opinion of physicians in general, whom he considered to be "ignorant of the known laws of electricity."[8] Volta would set his mind to the phenomenon without paying Galvani much heed.

Figure 4–3. Alessandro Volta. Courtesy Burndy Library, Norwalk, Connecticut.

First he repeated Galvani's experiments and was surprised to find that they worked. A key piece of his equipment was a Leyden jar, which he used instead of the friction generator to give the frogs shocks.

Galvani had encouraged Volta to repeat his experiments with frogs' legs, but Volta decided to use live animals instead. Perhaps Volta the physicist had a revulsion for dead creatures, although this did not stop him from reporting that "it is very amusing to make a [headless] grasshopper sing" when he used electricity to stimulate its nerve circuits. Whatever the reason, this variation had not occurred to Galvani, the anatomist used to handling dead animals.

The decision to use live frogs set the scene for a stunning discovery, which led to an almost unimaginable explosion of experimentation with electricity and magnetism around the world. In fact, Volta's work would transform the nature of the civilized world. One cannot help wondering how the history of science, and of our modern civilization, would have differed had Volta followed in Galvani's footsteps and used dead frogs for his experiments.

It was while experimenting with a live frog that Volta found that it could be made to twitch not only when stimulated by electricity stored in the Leyden jar but also when touched by a circuit made *of two different metals*. He concluded that some form of weak animal electricity had to be present in the body of a frog, and when it was made to flow through an external circuit it somehow disturbed the natural flow and that caused the twitching.[9]

Volta's next experiments were on himself. He connected a piece of tin resting on the tip of his tongue to a silver spoon placed further back in his mouth and experienced an unpleasant taste. (You can feel the same unpleasant effect by touching silver fillings with a spoon plated with or made of some metal other than silver because of the electric current produced by the bimetallic circuit. Perhaps your dentist could help demonstrate this for you.)

So Volta realized that the twitching of the frogs' legs was produced not from animal power but by contact between a variety of metals. This quickly destroyed any notion of the existence of animal electricity, an idea that had begun to enter the mainstream of public consciousness, no doubt spurred on by Aldini's popular road show.

Volta began a binge of experimentation during which he recognized that it was the presence of a moist interface between different metals that was essential for producing electricity. Almost any moist surface would do. The frogs had nothing to do with any of this. When Volta understood this point he had essentially made one of the greatest discoveries in the history of science. This was the important insight that Galvani had missed while conducting his orchestra of frogs' legs with a brass rod touching a silver box. Up to then the wet interface had been provided by the body of the frog, but the presence of animal tissue was irrelevant. Once Volta saw the light, frogs all over Italy heaved a croak of relief.

Volta had discovered that a moist contact of *any* sort between two different metals would generate electricity, as long as the liquid used was not distilled water (which conducts no electricity).

He began to experiment to discover how much electricity could be produced by connecting various metals via a moist conductor and described this quantity as their electromotive force. To detect whether any electricity was present he had to give himself a slight shock, or taste it on his tongue, a technique widely used by early experimenters who had no devices to measure amounts of electricity flowing in circuits (although placing a compass needle inside a coil of wire carrying current in what was called a galvanometer did become widely used later).

In 1800, after eight years of intense experimentation, Volta made public his invention of the device that made him famous. (To this day we are reminded of him when we consider the voltage of electric power in our homes or of batteries used in portable electronic devices.) The voltaic cell consisted of alternate layers of silver and zinc separated by pieces of moist cardboard, a device he described as an "artificial electric organ." Anyone who touched a pile of about forty or fifty pairs of these disks would receive a sensation akin to touching an electric eel. Such piles could stand up to eight feet high. (As kids we used to make "batteries" by placing blotting paper moistened by spittle between alternate layers of silver- and copper-plated coins and measuring their voltage, or touching wires from each end of this crude voltaic cell to our tongues.)

By now Volta was fifty-five years old, and he decided to withdraw from research to spend time with his three sons, all born since his marriage at age forty-nine. He played no further role in exploiting his inventions.

Voltaic cells could be easily constructed, and they soon appeared in laboratories all over Europe and America. At last experimenters had plentiful sources of electricity to work with, even if they did not yet understand why this odd device could produce electricity at all, or what electricity was. Chemists began to use electricity generated by voltaic cells to extract metals from ores, and others accidentally discovered that wires from voltaic cells placed in water created gas bubbles. The bubbles contained oxygen and hydrogen. Humphrey Davy (1778–1829) would become famous for identifying and naming the gases released in such experiments. The task of identification would begin with sniffing the gases, and some of the products of his research made him quite ill. He was the first person to get high on nitrous oxide, which came to be used as an anesthetic. Davy went on to discover potassium and sodium, and before he died he succeeded in identifying a total of forty-seven new chemical elements.

Bigger and better piles (batteries) were built. Larger and larger electrical currents were sent through certain metals to deposit other metals on their surfaces. To this day sheets of roofing material made of galvanized iron, in which zinc is electrically deposited on iron to provide long-lasting protection against corrosion, are used worldwide. (My first canoe was made of galvanized iron.) The name of the material reminds us of the time when electricity was still called "galvanism."

Thanks to frogs, Galvani, and Volta, the age of discovery in the physical sciences was about to begin in earnest. In addition to a nearly endless series of practical uses that were to be found for electricity, curious people began to use the voltaic cell in the study of magnetism.

NOTES

1. Quoted by Percy Dunsheath, *Giants of Electricity.* (New York: Thomas Y. Crowell Co., 1967), p. 4.
2. Ibid., p. 30.
3. We will meet voltaic columns in due course.
4. Dunsheath, *Giants of Electricity,* p. 50.

5. From his own report, "De viribus eletricitatis in motu musculari commentarius" (1791). See Galvani, *Dictionary of Scientific Biography*. (New York: Charles Scribner's Sons, 1970–1980).

6. Dunsheath, *Giants of Electricity*, p. 32.

7. Ibid., p. 107.

8. Volta, *Dictionary of Scientific Biography*.

9. Ibid.

n 5 n
Oersted and Ampère: The Birth of Electromagnetism

His constancy in the pursuit of his subject, both by reasoning and experiment, was well rewarded . . . by the discovery of a fact of which not a single person beside himself
had the slightest suspicion.

A description of Hans Christian Oersted,
Oersted and the Discovery of Electromagnetism

CURIOUS minds now began to make major break-throughs in understanding magnetism that would, at first, be based on keen perception of unexpected phenomena in the lab-oratory. Not all the minds involved were ready to see what serendipity placed before them, however. For example, during the twenty years following the invention in 1800 of the voltaic cell, no experimenter recognized the magnetic effects associated with electric currents.[1] An exception was a French chemist, Nicholas Gautherot, who in 1801 noticed that two wires connected to the ends of a voltaic cell tended to adhere to one another. But neither he nor any of the others who later reported that they had seen this phenomenon paid any further attention to it.

Louis Pasteur (1822–1895) once said, "In the fields of ob-servation, chance favors only the prepared mind."[2] In the an-nals of science we find that major discoveries are made when either nature or circumstance created by an experimenter pre-sents to his or her gaze a phenomenon that contains within it a major key to further insight. Such opportunities are not always accepted, nor even recognized. Usually the mind is prepared to recognize only what it expects and is therefore not open to the unexpected. This makes the task of finding a way through the unknown an adventure. A hazard along the path of progress is the presence of many apparently helpful signposts that lead nowhere. Some provide meaningful advice in the form of clues about which turn to take. Others lead to dead ends. Seldom is the path clearly marked. The scientist who successfully follows the trail left by the confusion of nature's clues is the one to whom history will later pay homage.

One mind that was prepared to take heed of a mysterious new effect was Hans Christian Oersted (1777–1851) (Fig. 5–1). Before he was forty-three, he too stamped his influence on history, in this case in a most remarkable manner. His key experiment was inadvertently performed in front of an audience.

Oersted was the eldest son of a poor apothecary, born in Rudköbing on the Baltic island of Langeland, Denmark, on August 14, 1777. He loved lecturing, especially as a popularizer of science, and he had done so ever since 1800, when he first heard about the invention of the voltaic cell and began to experiment with electricity by sending current through acids and alkalis. In those days he managed an apothecary shop for a professor who was on leave for a year and filled in for that gentleman at the University of Copenhagen. After making a favorable impression on all concerned, he received a travel grant from the university that took him to many laboratories throughout Europe. Upon his return he was given a small allowance from the state that allowed him to continue with his research. In 1806 he became professor of physics at the university.

More than a decade passed before the time was ripe for Oersted to step into history. The context from which he did so was remarkable. Since the 1780s it had been widely believed that Coulomb's research had shown that electricity and magnetism were in fact two "different species of matter whose laws of action were mathematically similar but whose natures were fundamentally different."[3] Oersted, however, believed that magnetism existed in all bodies and therefore had to be as general as electrical forces.[4] Also, he was aware that the Frenchman Ampère (whom we shall meet later) had announced in 1802

Figure 5–1. Hans Christian Oersted. Painting by D. Hvidt, after C. V. Eckersberg, 1822. Courtesy Burndy Library, Norwalk, Connecticut.

that electrical and magnetic phenomena were due to two different kinds of fluids that acted independently of one another.[5]

Many physicists regarded the apparent similarities between electric and magnetic forces as no more than an interesting coincidence. Oersted suspected that there was more to that similarity than met the eye, largely because of what he had read in the works of Immanuel Kant (1724–1804). That great philosopher, who might even have pleased William Gilbert, suggested that "science was not merely the discovery of nature; that is the scientist did not just record empirical facts and sum them up in mathematical formulas. Rather, the human mind imposed pattern upon perceptions; and the patterns were scientific laws."[6] Those patterns, Kant believed, were not arbitrary but lay rooted in the existence of Reason. He proclaimed his faith "that in reality there exists an underlying unity of the force of nature."[7] An instinctive awareness that nature is permeated by such unity played a profound role in the lives of several of the scientists we are yet to meet.

As Oersted later wrote about himself, "Throughout his . . . career, he adhered to the opinion, that the magnetic effects are produced by the same powers as the electrical. He was not so much led to this, by the reasons commonly alleged for this opinion, as by the philosophical principle, that all phenomena are produced by the same original power."[8] To him patterns in nature, or apparent coincidences connecting phenomena, indicated the existence of something deeper that was worth investigating. For example, his goal was to find under what conditions a conversion from electricity to magnetism, or vice versa, might take place. His initial tests led up an otherwise blind alley in which he discovered that a current sent through a thin wire made the wire hot. When the wire was made thin enough it also emitted light. He then incorrectly concluded that if he could make the wire even thinner—an impossible task at the time— magnetism would be created. That was in 1813. Oersted made no further headway with this line of research, at least not until a lucky day in 1820. But more than luck was involved.

Pasteur has stressed that chance favors the prepared mind. Oersted's was prepared, although the opportunity presented to him was nearly overlooked. He was still seeking a link between electricity and magnetism and thanks to the writings of Kant he really *believed* that such a connection must exist. Oersted was on a quest for the Holy Grail, which for him was represented by the connection between electricity and magnetism. Even if no one knew ahead of time how the link might be manifested, his mind was prepared to comprehend the slightest clue that might become evident.

When studying subjects like electricity and magnetism in

school it is rare that we hear about the sense of wonder that must have been felt by the pioneers as they struggled at the frontiers of knowledge, each explorer driven by some highly personal vision of what might be discovered in the laboratory. History only immortalizes those who made the "correct" discoveries, a judgment left for future generations to make. But for every "famous" pioneer in whatever field of science, there were hundreds of worthy individuals whose experimentation apparently led nowhere. Yet their work helped lay the foundations of their science and they, too, experienced the thrill of pursuing their research goals. This is worth bearing in mind as we follow the trail blazed by the experimenters who made the most important discoveries. Their work was performed in the context of a vast array of experiments by contemporaries whose names are forgotten by all except the most fastidious historians of science. The point of this diversion is to remind ourselves that whether or not a given experimenter takes a giant leap for humankind so often depends on luck befalling the prepared mind.

And so it was late in the winter of 1820, a year that he would later describe as the happiest in his scientific life, that Oersted gave his famous lecture. He began by stating that there had to be a connection between electrical and magnetic phenomena, a link that had been hinted at by a freak of nature. Lightning striking ships sometimes caused the polarity of compass needles to be reversed. No one understood why. In the days when compass needles were magnetized by rubbing them with lodestone, remagnetization of such needles was usually accomplished without further ado.

Oersted was also of the opinion that the magnetic effect of a current would only be manifested by using a very thin wire, which he had already shown became incandescent when a current was passed through it. At that time his conception of the nature of electricity involved the idea of "conflict." He believed that electricity did not flow in a uniform stream but in fits and starts, as if some aspect of it were in a state of perpetual conflict. He reasoned that when a current heated a wire or caused it to emit light, then, thanks to this conflict, both the heat and light radiated into space around the wire. Similarly, magnetic effects should radiate away from the wire under the influence of this clash of force. He planned to do an experiment that involved sending electricity through a thin platinum wire to see if it affected a compass placed near it. It was a simple test but he had insufficient time to try it before the lecture. He did not want to make an idiot of himself in front of an audience if the experiment failed and so decided to defer the demonstration.

During the lecture he came back to the idea of the relationship between electricity and magnetism and could not resist the temptation to perform his test there and then (Fig. 5–2). As a result the world's first demonstration of the bond between electricity and magnetism occurred before witnesses.[9] A compass was lying beneath a wire, and when Oersted turned on the current the needle deflected slightly.

Few in the audience were impressed. Neither was Oersted. The effect was so weak and unexpected, and he knew of others who had been confused by similar elusive phenomena, that he was not immediately convinced of its significance. He wrote later that he postponed further investigation of the phenomenon to a time when he hoped to have more leisure.[10] He went on:

> It may appear strange, that the discoverer made no further experiments upon the subject for three months; he himself finds it difficult enough to conceive it.[11]

With the assistance of several colleagues he resumed his experiments in July. These experiments have been described by science historian George Sarton as "among the most memorable in the whole history of science."[12]

What Oersted had actually observed, and then confirmed, was that a current flowing past a compass needle caused the needle to deflect. He found that a thicker wire produced a much stronger and unmistakable effect, and substances placed between the wire and the compass needle did nothing to interfere with it. From then on there was no holding him back. On July 21, 1820, he sent a four-page report[13] of his discovery to numerous scientific journals, and within weeks the news had spread far and wide. Oersted had proven that an electrical current could generate magnetism.

Other researchers had previously attempted this experiment, but they had begun by placing the compass needle at

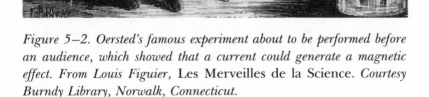

Figure 5–2. Oersted's famous experiment about to be performed before an audience, which showed that a current could generate a magnetic effect. From Louis Figuier, Les Merveilles de la Science. *Courtesy Burndy Library, Norwalk, Connecticut.*

right angles to the wire. No effect had been observed. Intuitively they expected that the magnetism created by the current should act in the direction of the current. In that case the needle should swing parallel to the wire. But nature pays no heed to our expectations. The magnetic force produced by a current is directed at right angles to the current, which meant that a compass needle would swing away from the axis of the wire. Although Oersted noticed a deflection of the compass during his public experiment, he had not neutralized the effect of the earth's pull on the needle. The needle could move through only a small angle before the earth's pull balanced the effect produced by the current. So in his experiment the needle did not end up pointing at right angles to the wire.[14] Nevertheless, he had observed the effect and that was enough to etch his name in history.

His was an accidental discovery and an incomplete breakthrough. He was looking for an effect and he could not have predicted that the needle would respond in the way it did. What became apparent to him, though, was that "the magnetical effect of an electrical current has a circular motion around it."[15] This was an awesome insight. Magnetism produced by a current did not act in straight lines, as everyone involved in the study of electricity and magnetism had assumed up to that time. Instead, it pulled the compass needle sideways with respect to the axis of the wire. This remarkable discovery of the difference between straight and circular magnetic effects removed a great conceptual block to progress.

Michael Faraday would later comment about Oersted's marvelous discovery, "His constancy in the pursuit of his subject, both by reasoning and experiment, was well rewarded . . . by the discovery of a fact of which not a single person beside himself had the slightest suspicion; but which, when once known, instantly drew the attention of all who were able to appreciate its importance and value."[16] In this manner the scene was set for the next great step in the understanding of both electricity and magnetism, a step taken by André-Marie Ampère (1775–1836) in France (Fig. 5–3).

Like so many of the pioneers traveling at the frontiers of knowledge, what Ampère discovered and how he interpreted his insights were largely determined by factors that had little or nothing to do with the subject at hand. Rather, it was his personal philosophy on life that cast the illumination over his search.

When we consider who made the necessary breakthroughs, which would eventually create for us the highly technological world in which we now live, it is worth looking at the role played by personal beliefs and expectations in the lives of some of those pioneers. Ampère grew up in a deeply religious atmosphere and was forever torn between his beliefs and what he learned

Figure 5–3. André-Marie Ampère. Courtesy Burndy Library, Norwalk, Connecticut.

from experience about the nature of the real world. He was born January 20, 1775, near Lyons in the village of Polymieux, son of a silk merchant of independent means who believed that a good education could be had by exposing the child to a good library and allowing him to choose his own course through it. Ampère flourished in this atmosphere and began to learn about the nature of the world as outlined in the thirty-volume Diderot encyclopedia. He studied at home and at his own speed. Well into old age he was able to recall long segments of what he learned during that unusual childhood.

All went well until his father was guillotined during the French Revolution. Now harsh reality confronted the eighteen-year-old head-on and he withdrew from social contact as he tried to comprehend the meaning of this needless death. During this vulnerable time he met Julia Carron, a somewhat older woman from a nearby village and significantly different social background who those close to him regarded as beneath his standing. He became infatuated with her and after considerable pursuit convinced her to marry him.

For the next four years, during which he became professor of physics and chemistry at a small school, he was happy. But then Julie died and Ampère was devastated. His personal life became a long-running catastrophe and his mental state caused his friends grave concern. Things became worse when his father-in-law from a second marriage, itself a disaster, swindled him out of what little money he had. Within two years Ampère was divorced. Yet his professional career flourished as he moved to ever more prestigious institutions. In 1824 he was elected to the chair of experimental physics at the Collège de France.

Despite his personal misfortune, Ampère made great strides in the study of electricity and magnetism. Initially his theories about these phenomena met with hostility and severe criticism. He persisted in his work, largely driven by an obsession about his claim to fame, alarmed at the notion that he might be overlooked. This obsession would obscure his ability to perceive what was revealed to him in his experiments, a perfect contrast to what we shall learn later about Michael Faraday.

Ampère had many preconceived notions about the subject of his studies. "Whenever he learned of a new effect in magnetism or electrodynamics, Ampère immediately tried to show that it was explicable in terms of motions of electrical currents."[17] This obsession acted as an impediment to being able to see things more clearly.

Nevertheless, "Ampère's personal misery had an important effect on his intellectual development."[18] Tossed back and forth between a world in which he sought truth and stability in the unearthing of the laws of science, as opposed to the "real" world in which chaos and confusion appeared to reign supreme, he also struggled to reconcile his strong religious faith with reality. He always sought certainty, and that search determined the evolution of his scientific work. For example, for Ampère it was important to accept both the existence of God and the existence of the world of external reality with which his scientific questioning had forced him into head-to-head contact. The two seemed irreconcilable, one rooted in faith, the other in fact. His way out of the dilemma was to find inspiration in the writings of Kant, whose philosophy made it possible to retain a religious faith in the context of living in a universe that appeared to function rationally. This helped Ampère deal with the inconsistencies of life.

Only when he had convinced himself that he could maintain a belief in physical reality and God was the stage set for Ampère to explore what could be learned about the former. He studied many areas of science, but it is what he learned about electricity and magnetism that enriches our story. Between 1820 and 1827 he founded the science of electrodynamics, the study of electricity in motion.

This chapter in his life began on September 4, 1820. Ampère was in the audience at a meeting of the Académie des Sciences in Paris when François Arago (1786–1853) reported Oersted's discovery that electricity created magnetic effects surrounding a wire carrying a current. This was stunning news, because everyone in the audience was aware of Coulomb's claim made in the 1780s that electricity and magnetism were not related. Unlike much of the skeptical audience, however, Ampère, now professor of mathematics at the École Polytechnique, immediately began to think about the implications of the new discovery.

Most researchers remain firmly attached to the exploration of widely accepted ideas. Although one does not run the risk of severe criticism, one is also unlikely to make significant strides in understanding. By definition, major advances are made in the face of current understanding (or dogma?) and are often treated the same way the body rejects a virus. A virus contains information in its molecular structure that the body must resist to survive unchanged. A new idea is rejected by the mind to preserve what it believes to be true. In this manner the mental immune system prevents infection by heresy.

In science it is taken as an act of scholarship to be able to show how well you can attack and demolish new ideas as quickly and efficiently as possible, especially when such notions do not instantly accord with generally accepted prejudices (or knowledge). This syndrome is one of the strengths of science. That is the good news. The bad news is that it creates a self-perpetuating system of scholarship that makes it difficult for new ideas to take hold. This is beneficial, as it prevents wild ideas from distorting orderly progress, but also restrictive, because creativity is kept under a very tight rein, sometimes to the point where it is not only scorned but taken as a sure sign of instability in the creative individual. In exceptional cases a new idea may be rapidly assimilated because the time is right for its acceptance, especially in the early days of a new discipline. But most often, new ideas, or even dramatic experimental breakthroughs, no matter how correct, have to suffer the ravages of time before they become widely understood or accepted.[19]

In beginning to play with a new idea about the relationship between electricity and magnetism, Ampère was willing to accept as possible the incredible, although, as we shall see later, that was not always so, not even for him. What he had heard that day at the meeting of French scientists was unbelievable: despite what Coulomb had claimed decades before, and in the face of what everyone else had believed since then, electricity and magnetism were related. So, after Arago's lecture was over and while the others stayed to argue about Oersted's claims, Ampère hurried back to his laboratory to begin his own experiments. Within two weeks he announced his own discoveries to the scientific world.

Ampère began by accepting the report of Oersted's work and allowed his imagination to play with ideas. He first repeated Oersted's experiment of placing a compass beneath a current-carrying wire and refined the set-up by neutralizing the influence of the earth's magnetic pull. This was done by placing magnets in suitable locations so their net effect precisely balanced and canceled out the influence of the earth's magnetic force through the space where his apparatus was mounted. Consequently, the influence of the current became easier to observe and allowed him to notice that the compass needle was deflected until it came to rest at right angles to the wire.[20] He also found that the current caused the compass needle to point one way beneath the wire and the opposite way above the wire. To him it was immediately obvious that this meant that the magnetic force formed a circle in space, concentric about the wire, as Oersted had already suspected.

Once he visualized this he imagined what would happen if he wound a coil of wire around a glass tube, for example. The magnetic force should then emerge from one end of the tube and enter the other so that the coil resembled a bar magnet. He tried the experiment and it worked. An iron rod placed inside a tube behaved like a bar magnet when the current was turned on. Was it possible, he wondered next, that terrestrial magnetism could also be accounted for in this way? Were circular currents running around inside the earth? Of course; that was it! That had to be why the earth was magnetic.

He next experimented with parallel current-carrying wires and found that they either attracted or repelled one another, depending on whether the currents flowed in the same or opposite directions.[21] This allowed him to formulate what was later called Ampère's Law, which describes the force between the wires. On September 25, 1820, he announced to the Académie that two helical coils of wire fed by a current (driven by voltaic cells, of course) could be made to attract or repel one another. Such coils form the heart of modern solenoids or electromagnets, used in automobile starter motors and relays, which operate everything from switches in electrical utilities to those that control air conditioning and heating units (at least until the advent of electronic chips).

The commercial exploitation of solenoids did not cross Ampère's mind as he sought to understand nature's secrets, even if "ideas flashed through his mind with such rapidity that he barely had time to note them down on scraps of paper before he set off chasing new ones."[22] He made his momentous discoveries despite the fact he was "dreadfully confused and . . . equally unskilled as an experimenter and as a debater."[23] So wrote Oersted after meeting Ampère. He found that the Frenchman was unable to explain his thoughts very clearly, could not understand the arguments of others, and, according to Oersted, most of Ampère's experiments didn't even succeed.

Ampère's greatest legacy was that he gave birth to the notion that *magnetism was produced by electricity in motion.* The compass needle placed beneath the current-carrying wire was in fact a detector of the current (current is electricity in motion). He had therefore invented a device for detecting current and called it a galvanometer.[24] It would become one of the most important measuring instruments for research into the nature of electricity and magnetism. With this he quickly proved that electricity actually flowed right through a voltaic pile, another major discovery on his part.

Ampere also stumbled onto the answer to the question we posed at the beginning of our book: "What is magnetism?" His answer was that magnetism was the force produced by electricity in motion. Herein lay the solution to the mystery of the lodestone and that of terrestrial magnetism. The connection to the earth's magnetism was made when Ampère realized that for the earth to be magnetic, circular currents had to flow beneath its surface. (This remains the essence of modern theories that account for the earth's magnetic field, invoking what is called the dynamo effect.) His experimental success was to reduce the problem of magnetism to that of electricity, although no one yet knew why tiny electric currents flowed in circles, either inside a magnet, inside a lodestone, or deep in the earth's core. Through a gem of insight he also explained magnetic polarity. "There is nothing more in one pole of a magnet than in another; the sole difference between them is that one is to the left and the other is to the right of the electric current which give the magnetic properties to the steel."[25]

Ampère was still interested in finding what lay at the deepest levels of nature. Inevitably he asked what gave rise to currents inside permanent magnets. Since he knew how voltaic piles operated, he considered that contact between molecules in the bar magnet must function as the contact between different metals in the voltaic pile. But a friend of Ampère's, Augustin-Jean Fresnel (1788–1827), the creator of the wave theory of light, pointed out that the idea could not be right because such current flowing in a poor conductor (like iron) would surely heat the metal and that wasn't observed. Fresnel was another creative individual, not saddled with tradition, and he suggested to Ampère that since the metal was made of molecules and nothing was known about molecules, why not imagine concentric electric currents flowing around the molecules themselves. Under the influence of magnetism, the molecules would be forced into alignment, after which the metal would exhibit its own magnetism. Ampère took to the idea in an instant. The theory he developed to account for this phenomenon worked so well that he used it to deduce Coulomb's inverse square law of magnetic attraction.

By now Ampère had unified the theories of electricity and magnetism. He recognized that they were connected at a level that remained beyond anyone's ability to observe directly (a level called the noumenal by Kant). As history would judge, his key idea, although intrinsically sound, was submerged in artificial notions about electricity, which he clung to because of what he believed about the nature of electrical currents in the first place. Together with many other physicists of his time, he believed that currents consisted of two fluids flowing in opposite directions. In the process these currents formed a "luminiferous ether" (or light-carrying medium) that pervaded space so as to support the two flows. That ether would quickly decompose as the electricity flowed through it. This made an explanation of the alignment of molecules in a magnet a very complex affair and we cannot dwell on the details. Today the explanation for currents involves the flow of electrons, which were not discovered until the early twentieth century, and magnetism works well without having to postulate oppositely flowing fluids or an ether.[26]

Despite his lack of understanding of the deep, underlying dynamics of the phenomenon of magnetism, Ampère was able to work out the details of his electrodynamic theory, which he set on a sound mathematical footing, but only after he had convinced himself as to what was taking place beyond awareness. To him the phenomena of electricity and magnetism behaved *as if* his model were true. This is often the case in science and shows that even incorrect theories can aid the mind in its search for truth. Ampère was obsessed with finding truth, and to him it was necessary to found his theory on something he considered to be understood in order to proceed with his research.

In 1824 he was given the chair of experimental physics at the Collège de France, a position he filled in addition to his role as inspector general of the university system, a post he had held since 1808 except for a few years in the 1820s. His magnificent accomplishments were complete. Then, driven by years of anxiety about money and his daughter's well-being in her continuing marriage to a drunkard, his health began to decline rapidly. In 1836 he died alone during an inspection visit to Marseilles as a result of complications from pneumonia.

In review, Oersted discovered that an electric current influenced a magnet, and Ampère showed that a current could actually *produce* magnetism. In the process he helped uncover one of the three laws of electromagnetism. The first law (Oersted's) stated that an electric current generated a field that was *circular* about the flow of current. The second law (Ampère's) stated that when currents flowed in parallel wires they attracted one another if the currents were in the same direction and repelled each other if the currents flowed in opposite directions. The third law (Coulomb's) was that the strength of the force between the current-carrying wires was proportional to the product of the strength of the current and inversely proportional to the square of the distance between the wires. An inverse square law of force also applied to gravitational attraction between masses, and it was not long before physicists began to search for the underlying phenomena that would link gravity and electromagnetism, a search that continues to this day. Physicists still adhere to Kant's intuitive feeling about the unity of the underlying forces of nature.

The next phase of progress was thus underway. Initial experimentation had led to the discovery that magnetism was created by electricity in motion, but the question of why magnetism could extend its influence over a distance with no obvious means of support still had to be explained. Further progress would require the physical insight of an untrained genius in England.

NOTES

1. Bern Dibner, *Oersted and the Discovery of Electromagnetism.* (New York: Blaisdell, 1962), p. 16.

2. Royston M. Roberts, *Serendipity: Accidental Discoveries in Science.* (New York: John Wiley & Sons 1989), p. x. This delightful book contains many stories related to accidental or serendipitous discoveries in science.

3. Hans Christian Oersted, *Dictionary of Scientific Biography,* (New York: Charles Scribner's Sons, 1970–1980).

4. From H. C. Oersted quoted by Robert C. Stauffer, "Speculation and Experiment in the Background of Oersted's Discovery of Electromagnetism." *ISIS* 48 (1957): 33.

5. Ibid.

6. Oersted, *Dictionary of Scientific Biography.*

7. Stauffer, "Speculation."

8. Oersted, *Dictionary of Scientific Biography.*

9. Stauffer, "Speculation."

10. Ibid.

11. Ibid.

12. George Sarton, "The Foundations of Electromagnetism." *ISIS* 10 (1928).

13. The text of the report is given in *ISIS* 10 (1928), and also by Dibner, *Oersted and the Discovery of Electromagnetism.*

14. L. Pearce Williams, "What Were Ampère's Earliest Discoveries in Electrodynamics." *ISIS* 74 (1983): 492.

15. Stauffer, "Speculation."

16. Quoted in Dibner, *Oersted and the Discovery of Electromagnetism,* p. 89.

17. Ibid.

18. Ampère, *Dictionary of Scientific Biography.*

19. This is an additional aspect of the phenomenon discussed by Thomas Kuhn in *The Structure of Scientific Revolutions* (University of Chicago Press, 1970).

20. Williams, "Ampère's Earliest Discoveries."

21. Arguments about how Ampère came to draw his conclusion are discussed by Williams, "Ampère's Earliest Discoveries."

22. Ibid., 494

23. Ibid.

24. Ibid., 499.

25. Ampère, *Dictionary of Scientific Biography.*

26. The Bohr model of the atom, which hypothesizes electrons or-

biting around the nucleus, provides all the moving electrons one wants, together with the subsequent creation of what is now termed a "magnetic moment," which is related to a quantity called the spin of the electron. This contributes to the phenomenon of permanent magnetism. But all that lay a century in the future.

∩ 6 ∩
Michael Faraday:
The Era of Discovery Personified

Faraday is the epitome of what can be accomplished by
self-tuition and enthusiasm, in spite of the most unpropi-
tious circumstances.
R. A. R. Tricker, *The Contributions of Faraday and
Maxwell to Electrical Science*

M ICHAEL Faraday (1791–
1867) was one of the great-
est experimental scientists who ever lived (Fig. 6–1). His role in
discovering more about the relationship between electricity and
magnetism highlights the next phase of progress. It was stimulated by his first conceptual insights into the nature of magnetism (and electricity) combined with clever experimentation.

Born September 22, 1791, in a slum south of London, the
third of four children in a poor family, Michael had virtually
no formal education. His story of success is one to inspire even
the most cynical.

Given his inauspicious start in life, no one could have foreseen that Faraday would emerge a great scientist, especially because, in his time, freedom of movement between careers was
next to impossible. His self-education rested on what he learned
while working for a bookbinder in an age when newspapers
were rented out. Faraday delivered these to clients, and those
that were not sold he read voraciously. Here we find a similarity
to Ampère, another avid reader and self-educated man.

At age nineteen, while still a bookbinder's apprentice, he

Figure 6–1. Michael Faraday. Courtesy The Royal Institution, London.

attended meetings of the City Philosophical Society in London, and it was there that he saw his first voltaic pile in operation.[1] Soon he built his own.

Faraday was excited by chemistry, but the opportunity to

become a scientist appeared to be nonexistent, especially because he lacked any formal education. At age twenty-one, just before he reached the end of his apprenticeship, a customer of the shop invited Faraday to join him in attending a series of lectures on chemistry to be given at the Royal Institution by Humphry Davy (1778–1829). Thrilled by the experience, Faraday organized the notes he had made of the lectures and sent them to Davy together with a letter seeking a post as his assistant. The notes deeply impressed Davy, but he was unable to hire the fledgling scientist until fate intervened. Davy was injured in a chemical explosion, and he asked Faraday to be his unpaid secretary to help him during the difficult period of recuperation.

The Royal Institution was, and remains, a remarkable organization. It was founded by Count Rumford (1753–1814), an American who earned a title in Bavaria before settling in England. The goal of the Royal Institution was to help spread knowledge and to introduce useful mechanical inventions and improvements into society. It was also a place for teaching "the application of science to the common purposes of life."[2] This was to become the place where Faraday taught and lived for the rest of his life, and where he carried out his research. His background completely closed the doors of academia to him, so he could not have moved ahead had it not been for this remarkable institution. During the 1820s, Faraday helped raise money for the Royal Institution by performing chemical analyses and organizing and giving public lectures, sometimes before royalty. His popular presentations on the candle flame became classics of their genre.

In March, 1813, a few months after Faraday began to serve Davy as a volunteer, a laboratory assistant was fired for brawling. Davy lobbied to the managers of the institution to allow Faraday to fill the position and said that "his habits seem good, his disposition active and cheerful, and his manner intelligent."[3] The managers were persuaded and Davy sent news to Faraday. The young man quit his bookbinding career on the spot.

In his new role as a professional chemist, Faraday began to delve into the nature of matter with a view to understanding how various substances could chemically combine. Electrolysis, the decomposition of liquids that conducted electricity, played an important role in the early evolution of his ideas. Electrolysis had been discovered by accident on May 2, 1800, when the first large voltaic pile was built in England. To insure a good contact where a conductor touched the upper plate, researchers had placed a drop of water on it. When electricity flowed the water bubbled, which meant it was producing gas. Further experiments were carried out by placing two platinum wires connected to a voltaic pile in water; they created hydrogen and oxygen at the two terminals.

These discoveries led to the notion that "galvanic" or current electricity actually flowed through a circuit, as opposed to remaining static. Faraday began working with Davy on the study of electrolysis, and after their paths separated years later Davy went on to identify forty-seven new elements produced using the techniques of electrolysis. Faraday would continue to experiment with electricity and magnetism.

Meanwhile, however, Faraday's education benefited enormously when he accompanied Davy on a tour through Europe that took the better part of two years. During the trip he met many distinguished scientists and was able to learn from them at first hand.

Until 1821 his interest in electricity smoldered in the background because of the pressure of other duties. But he kept himself informed as to what physicists were thinking at the time. The widely accepted notion was that electricity involved central forces emanating from particles and that such forces acted in straight lines,[4] without any obvious means of support, to touch nearby objects. This "action at a distance"[5] was believed to occur instantaneously. The formulas describing the force were given in terms of the properties of the matter involved in producing the force and the distance between two masses or charges.[6] There was nothing in the equations to suggest that time was involved. With hindsight it is apparent that until the time element was considered, no further progress in the theoretical understanding of either electricity or magnetism was possible. But what would cause the human mind to take the giant step of introducing time into a description of a phenomenon that appeared to take place instantaneously?

How and why Faraday managed to do so was largely guided

by his personal approach to the quest, which was defined by his religious outlook on life. He was another of the major figures in our story who were imbued with the Kantian metaphysical belief in the unity of the forces of nature. But to him it was more than that. His religious upbringing, within an obscure Christian sect called the Sandemanians, led him to apply what he learned in his church to his studies of nature.

Sandemanians lived their lives according to the Bible and in imitation of Christ's perfect thoughts and deeds, but with a difference. There could be no intermediary between them and the Bible: in the Sandemanian sect there were no clerics to act as interpreters of the written word, the alleged source of truth. They taught that to find the truth all you had to do was look carefully and closely enough and you would see it for yourself. No one could help anyone else do this. Once truth was seen in this manner, the interpretation could not be argued. And that was how Faraday approached nature: he learned to confront nature *directly*. In that way he would learn her secrets, in particular those related to the unity of forces.

Faraday discovered nature's laws by direct observation and experiment, not by listening to what theoreticians said he should find. For him this left very little room for mathematics, which could not help anyone actually see more clearly. Mathematics might help others to describe what was found after the fact, but it appeared to serve no purpose in guiding the search any more than a cleric could help one discover what was written in the Bible. For Faraday the truths of nature were there to be observed directly with no intermediary (such as a college professor) to tell him how to look and where to seek. To make progress, all the dedicated researcher had to do was learn to see clearly. That Faraday was able to do better than almost anyone else in his time.

Faraday became an elder of the Sandemanian Church where, it was said, he preached badly. Lecturing, not preaching, was his forte. He was once removed from his office of elder, which required unfailing regular attendance at church, because he missed a meeting without adequate reason. It was insufficient cause that he had been commanded to dine with the Queen at Windsor Castle.[7]

To understand the magnitude of this remarkable genius's

mind, consider some of the opinions widely held in Faraday's time. Scientists argued about whether a vacuum was possible and whether two forms of matter, ponderable (influenced by gravity) and imponderable (not so influenced), might exist. It was believed that electric fluids were in the latter class. Static electricity and magnetism were both considered to involve the flow of one or two fluids. Usually, if two fluids were present they balanced each other and didn't flow anywhere. Current electricity, however, was produced when these fluids flowed. In the case of magnetism the fluid had to have magnetic properties.

Ampère believed that an electric current was sent on its way because of the polarization of particles in a compound that caused positive and negative particles to separate. These would then exchange places until the particles reached the end of the circuit. For example, starting with a set of unpolarized particles, not lined up, one could imagine the following random alignment in a wire:

$$+ -, \ - +, \ + -, \ + -, \ - +.$$

When a circuit was formed, the particles would first rearrange to be oriented like this:

$$+ -, \ + -, \ + -, \ + -, \ + -$$

Then the pairs of positives and negatives would exchange positions and after a number of steps you would have, in essence:

$$+ + + + + \qquad - - - - -$$

This mechanism was thought to explain currents and action at a distance.

After Oersted found evidence for a circular pattern in magnetic force around a current-carrying wire, most scientists continued to cling to the notion that straight forces emanated from central particles and that this accounted for observed phenomena. Faraday, because he was less steeped in tradition, was willing to explore an alternative: the existence of circular forces.[8] To him it seemed an elegant way to create a polar situation, with two magnetic poles connected by some axis surrounded by circular force. Herein lay the "germ of the idea of the line of force which was to be central to the development of Faraday's theories."[9]

News of Oersted's discovery had reached the Royal Institution on October 1, 1820, and soon scientific journals were overwhelmed by often confusing articles on the subject. In the late spring, early summer of 1821, "Richard Phillips, one of the editors of the Annals of Philosophy and a long time friend, suggested to Faraday that he write an account of what had been done up to that point in the new science of electromagnetism. It was this suggestion that started Faraday on his 40-year quest."[10] Phillips hoped that Faraday could sort through the confusion and determine which ideas had meaning and which did not.

While he was studying the subject, Faraday was also reviewing the work of some of the greatest scientists of his day, a mighty job for someone of such lowly beginnings. This situation caused him to write that he was "but a young man and without the name, and it probably does not matter much to science what becomes of me,"[11] a sentiment that reveals much about his self-image at the time.

When Faraday started this work he did not have proof that electricity was material, "or of the existence of any current through the wire."[12] There was no current-measuring device available at the time, although experimental phenomena were consistent with the notion that something was flowing in the wires. It appeared "as if" electricity was a fluid. Questions about what this mysterious fluid was were widely regarded as having no relevance.

At the start of his project, Faraday was aware that Ampère's theory contained five points, none of which pleased him.

1. Magnetism was the result of current electricity.
2. Permanent magnets contain circular, aligned (or coaxial) electrical currents around particles inside the magnet.
3. Electrical currents consisted of two fluids created by the breakdown of the luminiferous ether.
4. Attraction and repulsion between current-carrying wires resulted from *central* forces between the currents.

5. The central forces were propagated through the air by vibrations of a luminiferous ether.

A debate ensued between Faraday and Ampère, which has been discussed by science historian L. Pearce Williams,[13] who argued that Faraday would later push Ampère to become more critical of his own ideas by insisting on the primary importance of experimentation and the secondary importance of theory. This stimulus led Ampère to give up some of his ill-found notions with the result that he became freer to make his own significant contributions to the theoretical understanding of electricity and magnetism, contributions so significant that Maxwell would later label him the Newton of electricity, owing to Ampère's law of electrical attraction, which is similar to that of gravity.

At the start of his own experimentation, Faraday bore in mind that Oersted had described the effect of an electrical current stimulating a nearby magnet in terms of an electrical "conflict" that "performed circles" around the wire. He did not understand Oersted's theory of electrical conflict, of two fluids flowing in opposite directions, because that should cause opposite flows to cancel, unless one acted only on the north pole of a magnet and the other only on a south pole. That seemed artificial to Faraday. If it were so, what happened when the two fluids first met? Surely they couldn't easily slide past each other.

Faraday repeated Oersted's experiment and saw that when a compass was moved around the wire the poles rotated so that the magnetized needle was always tangential to a circle centered on the wire. Because he knew that the two poles were reacting equally while being pulled in opposite directions, it was logical to imagine that a *single* pole would rotate endlessly about the wire as long as the current flowed. Because single magnetic poles did not exist, he could not perform an experiment to test this idea. (The possible existence of magnetic monopoles remains a profound question that continues to annoy and fascinate modern cosmologists trying to explain the creation of the universe and physicists dealing with the fundamental forces of nature.)

Faraday devised a very ingenious device to demonstrate what Oersted had discovered, that a current-carrying wire did produce a circular magnetic force. In the absence of a magnetic

Faraday's rotation experiment

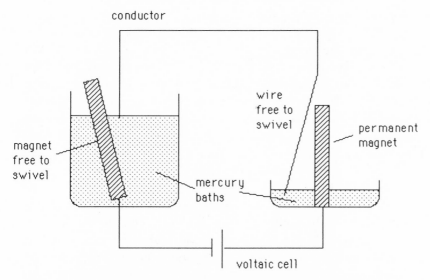

Figure 6–2. A sketch showing the apparatus Faraday used to demonstrate the mechanical effects of electric currents. When a current flowed in the circuit, the suspended magnet at the left rotated around the wire firmly fixed in the bowl of mercury. Simultaneously, the current-carrying wire at the right rotated about the permanent magnet anchored in another bowl of mercury.

monopole with which to perform the experiment, he set up a circuit in which a bowl of mercury was placed with a vertical magnet through it (Fig. 6–2). The electrical circuit was completed by having a wire loosely suspended in the mercury. This was one side of the device. On the other side a magnet was free to move around inside the mercury while the electrical conductor was fixed. When a current was sent through the device, the wire rotated about the magnet on the right-hand side and the magnet on the left-hand side rotated around the wire. The current caused mechanical effects.

On Christmas day in 1821, his brother-in-law, George Barnard, was there to see it happen. "I shall never forget the enthusiasm expressed in his face, and the sparkling in his eyes,"[14] Barnard wrote. Faraday's report of his seminal work was entitled "On Some New Electro-Magnetical Motions and on the Theory of Magnetism." By using electricity to make something move he had taken the first giant step toward the creation of the modern industrialized world, which depends so heavily on the use of the electric motor.

His report described the circular motion in terms of "lines of force" set up around magnets or current-carrying wires. The concept of lines of force still permeates all of physical science. The idea that these lines of force were circular was simply the outcome of his experiments. Faraday was not put off by the notion of circularity, unlike other physicists who continued to adhere to the idea that the relevant forces could act only in straight lines.

He proceeded with his experiments unhampered by crippling beliefs about the way things should be. But for his contemporaries there was nothing simple in any of this, since it was mathematically very difficult, if not impossible, to convert straight lines into circles.

In his endless search for truth and knowledge, Faraday knew that God would not make the task easy. He had a strong sense that he had to make sure, through proof and disproof, that what he said was correct; otherwise others would surely point out, by means of scientific proof, that he was incorrect.

Concerning an ability to consider new points of view, Faraday would later describe what a scientist should be, although he labeled that sort of person a natural philosopher.

> The philosopher should be a man [sic] willing to listen to every suggestion, but determined to judge for himself. He should not be biased by appearances; have no favorite hypothesis; be of no school; and in doctrine have no master. He should not be a respecter of person, but of things. Truth should be his primary object. If to these qualities he added industry, he may indeed hope to talk within the veil of the temple of nature.[15]

Faraday exhibited an outstanding ability to work hard and to keep his mind open to appreciate the lessons nature laid before him. He was willing to consider points of view that his colleagues in other laboratories avoided. Perhaps modern university curricula should include exercises to facilitate the letting

go of prejudice so as to encourage students to consider occa-
sionally the impossible if not the incredible. The time and again
significant breakthroughs in science are made by those ready
and willing to take a completely different approach to a problem that had been unsuccessfully confronted by more conventional minds for years. The creative approach requires that, for
a while at least, someone be willing to climb out of the mainstream and take a look at the flow of ideas from the banks of
the river. How else does one obtain a clear perspective? If you
are up to your neck in the water it is difficult to see where you
are, and impossible to perceive the world view that can be had
from the bank.

Faraday never even entered the mainstream. From his unique
perspective grew the idea of magnetic fields of force, an insight
that emerged while he was studying patterns made by iron filings sprinkled on pieces of paper laid over various combinations of magnets (Fig. 6–3). He had first noticed such patterns
when Davy "sprinkled iron filings on a sheet of card through
which a vertical wire was passed. The pattern produced suggested a structure of concentric rings in a plane perpendicular
to the current."[16] Davy was originally driven to trying these explorations because he did not understand what Oersted was
trying to say about circular forces. Faraday took up the quest
and began to map the magnetic pattern under various conditions, with magnets arranged in different configurations with
respect to one another. This led to his concept of a field around
a magnet in which the presence of the magnetism was manifested.

He next conjured up a beautifully simple experiment to discover how circular forces emerged from the center of a current-
carrying wire. Like Ampère, he reversed the problem. If circular fields emanated from straight wires, would a straight force
be created by a current flowing in a circle? To demonstrate this,
Faraday wound wire around a glass tube and half submerged it
in water in which floated a compass needle. When the current
was switched on, the needle swung to point to one end of the
tube and was pulled into it until it came to rest when its north
pole drew up to the north "pole" of the field produced inside
the tube. Faraday then realized that a monopole would move
endlessly in and out of the coil wound around the glass tube.

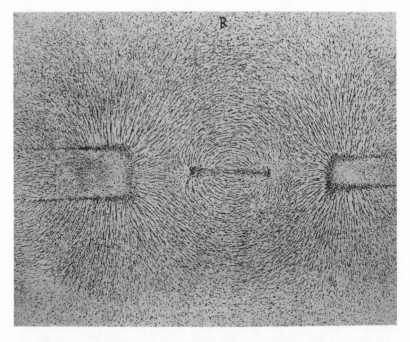

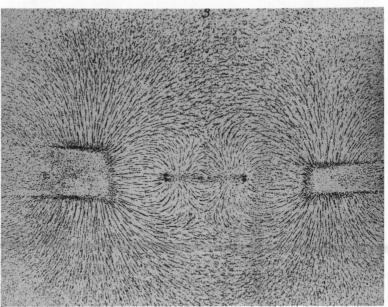

Figure 6–3. A pair of Faraday's sketches of the magnetic field patterns he observed when he sprinkled iron filings on paper laid over various combinations of magnets. Courtesy The Royal Institution, London.

The windings of the coil concentrated the lines of magnetic force inside the tube. Outside, the "lines of force" looked just like those he had found around a permanent magnet. The pattern was the same as Gilbert had discovered 220 years before when he moved a compass needle around the spherical lodestone and inferred, from the similarity of that pattern to the way compass needles tilted, that the earth was magnetic.

In September, 1821, while others continued to think of electricity in terms of fluids within substances, Faraday was off on his own, thinking about circular lines of force and wondering what caused these lines to come into existence. Something had to be transmitted through space around the wire, something that actually filled that space. Just as the phenomenon of magnetism in lodestones had mystified people for millennia, Faraday's work did little to remove the mystical or magical aspects of the basic phenomenon. He could, however, create magnetic magic by using electricity and could see, through his iron filing maps, how the lines of force outlined a field of influence around the magnet.

When his work was complete, he wrote "A Historical Sketch of Electromagnetism" and published it anonymously, perhaps reluctant to reveal that a relative amateur had dared to present such important discoveries to the world. In the book he outlined some principles that still guide experimentalists today: [17]

1. It is not scientifically proper to make up states or entities for which no experimental evidence exists.
2. Hypotheses cannot be freely invented, but must have some experimentally verifiable aspect.
3. Hypotheses must be clear and unambiguous, and they must serve to explain, in a mechanical way, the phenomena for which they were invented.

He was skeptical of theorists whose notions so quickly collapsed when new data came to light. To this day it remains a cliche in scientific circles that some of the theoretically inclined cannot deal with "reality," i.e., with experimental results. Faraday was the exact opposite of this. Rather than swallowing theory, he preferred "some facts to help me on." [18] These facts he had to provide for himself, as any good Sandemanian was expected to do.

A most important conceptual breakthrough, which would irreversibly alter life for every human being on our planet, now lay around the corner. Faraday asked whether a field induced by a current-carrying wire could, in turn, induce a current in another wire placed near it. This question may appear to be but a small step, but when he found the answer it was a gigantic leap for humankind. A decade would pass before he took that step.

Sometimes matters of logical deduction may, with hindsight, appear to have been trivial. Yet conclusions are only arrived at after the individual brain has had time to filter, mix, and redistribute ideas and insights. The concoction must then be allowed to marinate before it is ready to be sampled. In the words of Hercule Poirot, Agatha Christie's detective of action, to draw new conclusions "the little grey cells require time to do their work." Then, as if by inspiration from some metaphysical wellspring of being, an insight may emerge. The religiously inclined attribute such insights to a divine source, while the artist credits the muse. Scientists refer to this as inspiration, which often arrives unexpectedly—"out of the blue." Yet the underlying process is the same for all of us. The unconscious is fed a question. If it has access to relevant data and related ideas, new connections are made beyond awareness until, as if by magic, the solution presents itself. All one has to do to encourage the process is to pose the question, make available as much information as possible (by reading, study, talk, and experiment), and trust in the outcome. Of course, for the scientist any inspiration generated in this manner must be tested against reality to find whether it is relevant, whether it accords with nature's ways. This restriction does not slow down the artist whose inspired work of art is a personal statement that need not be tested against the laws of nature. Nor does the believer have to test the validity of an insight before using it to deliver a sermon to inspire the faithful (and in the process give thanks to some holy spirit for the insight in the first place).

From 1824 to 1831 Faraday tried to find a way to show that magnetism could produce electricity. He knew that electricity, i.e., the flow of current in a wire, produced a magnetic field

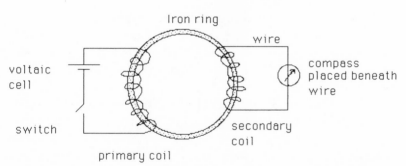

Figure 6–4. A sketch showing the elements of the apparatus Faraday used to discover magnetic induction. A voltaic cell at the left provided current that could be sent through a coil of wire wound around half of an iron ring. A second coil was wound around the right-hand half of the ring and its ends joined in such a way that the wire ran above a compass. When the switch was closed and current flowed into the primary coil, a magnetic field was generated in the iron ring. That field passed into the second coil and generated a brief surge of current, which was detected by a deflection of the compass needle.

around the wire. So, why wouldn't the corollary be true? Shouldn't a magnetic field produce a current?

During this time his mind had been going around in circles, which was the nature of the magnetic force he was thinking about. Then he heard that Joseph Henry, in the United States, had built a powerful electromagnet by winding a coil of wire around an iron bar, and that the reversal of the current instantaneously reversed the polarity of the magnet. Faraday's mind was ready. He imagined winding a coil around one side of an iron *ring*. Surely, whatever was involved in creating the magnetism in the iron would propagate through the entire ring. If he then wound a coil around the opposite side of the ring, a current might be produced in it as the result of the initial magnetization of the ring (Fig. 6–4).

In this circuit, when a current runs through the primary coil on the left, a steady magnetic field penetrates the iron. If the secondary coil wound around the ring sensed this field, any current produced should affect the compass needle placed beneath the wire that was part of the secondary circuit. Faraday switched on the current and watched what happened. Nothing; or at least nothing he had expected. When the current was switched on, the compass needle gave a little kick and then returned to its usual, motionless state. When the current was switched off the needle kicked again.

That, it turned out, was a stunning discovery. The compass needle only deflected at the moment the current in the primary was switched on or off. Faraday figured out that when the current was switched on it built up a magnetic field that penetrated the iron ring. The field created in the ring built up from zero to some final value and while it did so the field lines moved through the iron. That, in turn, made the entire ring magnetic and its field lines then literally moved through the secondary coil. Only when the field lines moved was a current produced. When the field reached its final configuration, determined by the steady current flow, the compass needle returned to its normal position. When the current in the primary was switched off, the field collapsed and disappeared and in the process of collapsing a current was momentarily triggered in the secondary circuit.

Faraday recognized that the existence of a changing magnetic field was the key to understanding the brief kick of the compass needle when the current was switched on or off. The needle responded to the field only as it built up and the lines of force literally cut through the wire of the second coil. Otherwise the needle settled back to rest. When the switch was left on, a steady current flowed into the primary and no response was seen in the secondary circuit. Nothing was changing; the magnetic field was in a steady state. Within an instant of switching the current off again the field collapsed and a brief surge of current was again detected.

Faraday's great insight was to recognize that the magnetic force had to be *changing* in order to produce a current. This was totally unexpected. After all, Ampère and Oersted had shown that a steady current created a steady field. Who would have imagined, except with hindsight, that a changing field was required to provide a current?

The phenomenon he observed came to be known as electromagnetic induction, or, simply, induction. The current in the first coil *induced* a field in the iron ring, which, in turn, induced a current in the secondary circuit.

This extraordinary discovery would loom large in the history of science. What he had discovered was that a magnetic field could be generated by a steady current, but a current could only be driven (or charge be made to flow) by a *changing* magnetic field.

It was in January, 1832, that Faraday "speculated that electromagnetic induction might take place through the 'cutting' of lines of force."[19] Two months later in a note to the Royal Society he wrote:

> When a magnet acts upon a distant magnet or piece of iron, the influencing cause . . . proceeds gradually from magnetic bodies, and requires time for its transmission which will probably be found to be very sensible.
>
> I think also that I see reason for supposing that electric induction (or tension) is also performed in a similar progressive way.[20]

Faraday had discovered that it was a *changing magnetic field* that generated the current. This meant that *time* had to be involved in any mathematical description of electricity and magnetism. Because induction required varying currents or changing fields, it meant that time had to be a factor in any explanation for the phenomenon. Up to then the forces were believed to propagate instantaneously. Now the possibility loomed large that a finite speed was involved. That speed was unknown.

He did not discover induction during his first bout of experimentation in 1822 "because he did not recognize the possibility that the expected effects could be transient."[21] It took nine years before his cognitive system became capable of recognizing the importance of a transient phenomenon. Once he recognized it though, it seemed so obvious! As a result of his discovery the way to deeper theoretical understanding was opened.

Faraday next asked how a *steady* current of electricity could be generated at the second coil. A few weeks later he realized that a *continuously moving magnetic field* would generate a steady current. So he made a copper disk rotate with its edge between

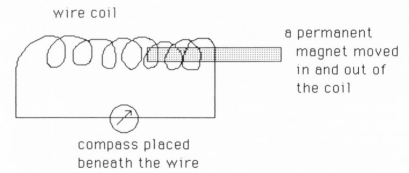

wire coil

a permanent
magnet moved
in and out of
the coil

compass placed
beneath the wire

Figure 6–5. A sketch illustrating how Faraday managed to produce a steady current, by moving a magnet into and out of a coil of wire the ends of which were joined to pass over a compass. Only while the magnet was moving was a current produced in the coil, and that was shown by the deflection of the the compass needle.

poles of a magnet and found that current flowed from the center to the edge of the disk (or vice versa). This was the world's first dynamo or electric generator. Then he moved a magnet in and out of a helix of wire to create a current (Fig. 6–5). A more efficient way to produce electricity was to move coils of wire in a field, the principle upon which electrical generators are built. They would soon relegate voltaic cells into the dusty closets of history.

Michael Faraday's magnificent insight into the nature of induction was that time was of the essence. His little gray cells had allowed him to recognize this; they had made the right connections in his mind. He was mentally prepared for this insight, thanks to other projects he had been working on immediately preceding the breakthrough. Those projects were unrelated experiments that specifically involved *transient* phenomena, so he was mentally primed to be aware of them, a state of mind essential for his recognition of how induction worked.

By now Faraday had survived several laboratory explosions, one of which left pieces of glass in his eyes. During the period 1831–1840, he overworked and, as one biographer put it, "The strain of eight years of unremitting intellectual effort at the farthest frontier of electrical theory ultimately broke his powerful mind."[22] In 1839 he suffered a nervous breakdown. His illness was marked by giddiness and memory loss, which "affected the working of his brain, though his body retained its full strength."[23] He never fully recovered. He may also have suffered from mercury poisoning: his body showed the symptoms and mercury was common in laboratories in his day.

"As his mental faculties declined, Faraday gracefully retreated from the world."[24] He kept on giving popular lectures until, in 1862, he retired to a house given to him by Queen Victoria in honor of his great achievements. He died on August 25, 1867, at the age of seventy-six.

In review, in 1821 Faraday discovered that electricity could produce mechanical action (as he demonstrated by having a wire move around a current-carrying conductor). In 1831 he found that mechanical movement of magnets could produce electricity. The scene was thus set for modern technological civilization to burst forth. These two principles would be exploited in the construction of electrical generators, in which moving magnets create current, and electric motors, in which changing currents are used to move magnets; that is, to cause mechanical motion. Together, the electric generator and motor are the source of power in modern society and drive the machines of industry.

NOTES

1. Michael Faraday, *Dictionary of Scientific Biography*, (New York: Charles Scribner's Sons, 1970–1980).

2. D. K. C. MacDonald, *Faraday, Maxwell, and Kelvin.* (New York: Anchor Books, Doubleday, 1964), p. 14.

3. Ibid., p. 19.

4. Faraday, *Dictionary of Scientific Biography.*

5. For a fascinating history of action at a distance in the nineteenth century see A. E. Woodruff, *ISIS* 53 (1962): 439.

6. For example, in the case of gravity, the archetypal example of force at a distance, the force between two masses is given by

$$F = G \frac{m_1 m_2}{r^2}$$

where G was the universal gravitational constant, m_1 and m_2 the masses of the two attracting particles, and r the distance between them.

7. R. A. R. Tricker. *The Contributions of Faraday and Maxwell to Electrical Science.* (Oxford: Pergamon Press, 1966), p. viii.

8. Faraday, *Dictionary of Scientific Biography.*

9. Ibid.

10. L. Pearce Williams, "Faraday and Ampère: A Critical Dialog." In *Faraday Rediscovered,* edited by David Gooding and Frank A. J. L. James. (New York: Stockton Press, 1985), p. 86.

11. Ibid., p. 89.

12. Ibid., p. 88.

13. Ibid., p. 83.

14. Quoted by Percy Dunsheath, *Giants of Electricity.* (New York: Thomas Y. Crowell Co., 1967), p. 108.

15. MacDonald, *Faraday, Maxwell, and Kelvin,* p. 51.

16. David Gooding, "In Nature's School." In *Faraday Rediscovered,* edited by David Gooding and Frank A. J. L. James. (New York: Stockton Press, 1985), p. 113.

17. Pearce Williams, "Faraday and Ampère," p. 88.

18. Ibid., p. 89.

19. Nancy J. Nersessian, "Faraday's Field Concept." In *Faraday Rediscovered,* edited by David Goody and Frank A. J. L. James (New York: Stockton Press, 1985), p. 182.

20. Ibid.

21. Ryan B. Tweney, "Faraday's Discovery of Induction: A Cognitive Approach." In *Faraday Rediscovered,* edited by David Goody and Frank A. J. L. James. (New York: Stockton Press, 1985), p. 198.

22. Faraday, *Dictionary of Scientific Biography.*

23. Tricker, *Contributions,* p. 18.

24. Faraday, *Dictionary of Scientific Biography.*

∩ 7 ∩
Fields and Faraday

The nation that controls magnetism controls the universe.
                                                    Dick Tracy

PROGRESS in the study of magnetism and electricity was always able to move forward unswayed by the caution of clerics or theology. Nowhere in any of the world's fundamental religious tracts were there statements about the nature of magnetism. This meant that physicists did not have to struggle against church doctrine, as was Galileo's lot when he tried to show that the sun was at the center of the solar system. No segment of the population was offended by discoveries about the nature of magnetism in the way that religious fundamentalists continue to rankle at the discovery of biological evolution. This freedom to explore beyond any possible restraints placed on pursuing certain other questions may account for the rapid progress in physics that took place in the nineteenth century. This is not to imply that the physicists involved did not hold religious beliefs. We have seen that many of them did; only that these beliefs did not hamper progress. In the case of Faraday, they helped.

Another aspect of scientific research that we should not forget is that in a story like ours we can touch upon the work done only by those who made what posterity would call the major breakthroughs. The names of many hundreds of other scientists whose research efforts, personal beliefs and expectations, or carefully worked out theories did not resonate with nature's truth have been lost to posterity, while the names of Oersted,

Ampère, and Faraday forever ring through time. Yet even these men on occasion followed the wrong clues or overlooked a new phenomenon and thus missed making even greater contributions to their science.

Ampère, whom we have already met, was a fascinating example. Although his personality and approach to science were ideally suited for taking major steps forward, his psychological state prevented him from making more progress than he did. He was guided by theoretical expectations to search for specific phenomena. These he *expected* to observe. On one occasion, when nature offered him something different, he could not see it. This occurred in 1822, nine years before Faraday discovered induction. Ampère performed a series of experiments that produced unexpected results that historians now realize were unequivocal evidence for induction, but to Ampère the effect, although noted, was simply ignored. It was not what he was looking for and he failed to recognize its significance.[1]

What our brains are capable of perceiving in the behavior of nature is heavily determined by what we are ready to see and, perhaps more importantly, by what we expect to see. This is a hazard all scientists confront, one to which most of them assume they are immune. But when experimenting at the borders of the unknown, in search of a particular phenomenon that might help cast light, our unconscious dependence on theories on how nature behaves makes the path of progress rocky.

Faraday was a scientist more likely to explore the unknown for the sake of exploration than to perform experiments to test theories. He was not only willing but uniquely able to proceed without any theory or beliefs in mind. For that reason Faraday became known for discovering electromagnetic induction, the same effect seen but overlooked by Ampère nine years before.

Meanwhile, though, after his breakdown, Faraday recovered sufficiently to resume his attempts to better understand electricity and magnetism. Although scientists were using electric currents, no one knew what electricity was. That mystery was just as profound as it would have appeared centuries before to William Gilbert or Peter Peregrinus if they had been asked to give a complete physical or mathematical description of electricity or magnetism. Since then, ever-increasing numbers of experiments had been performed and scientists had learned to create both electricity and magnetism at will. They had also shown that magnetism and electricity were related. Ancient questions about the mystery of lodestone could be answered. At first it was enough to say that it was due to magnetism. But what was magnetism? Ampère's answer was that magnetism was a force produced by electricity in motion. However, that still did not explain why magnets had the magical ability to reach out and influence objects located some distance away, nor why a lodestone was magnetic in the seeming absence of any electric current.

At a deeper level, then, no one knew what magnetism or electricity was, although Faraday showed that the field concept was a powerful tool for understanding magnetic phenomena. It was left to the mathematicians to take the idea and exploit it. The manipulation of fields as mathematical concepts was something that could be done readily by then.

On August 6, 1845, William Thomson (1824–1907) wrote a letter to Faraday explaining how he had been able to place the notion of fields on a solid mathematical foundation. According to Thomson (later to become Lord Kelvin, after whom the unit of absolute temperature, the degree Kelvin, is named), magnetic fields should also influence polarized light propagating through glass. Two decades before that, Faraday had done research on the structure of glass and had produced a piece with a very high refractive index. He had already tried to detect whether an electrical force applied to the glass might cause some unusual interaction with light passing through the glass, but had failed. Now he was ready for another look. Why had Faraday expected that light and magnetism were related? Because he had an intuitive feeling that all of nature's forces were somehow related, the same Kantian attitude reflected in the personal philosophies of Oersted and Ampère. As the British physicist John Tyndall (1820–1893) later wrote, Faraday "had views regarding the unity and convertibility of natural forces, certain ideas regarding the vibration of light and magnetic force; these views and ideas drove him to investigation."[2]

Motivated by Thomson's letter, Faraday returned to the laboratory on August 30, 1845, to search for an effect of electricity on light (Fig. 7–1). At first he shone light through electrolytes in which current was flowing. Nothing happened to the light.

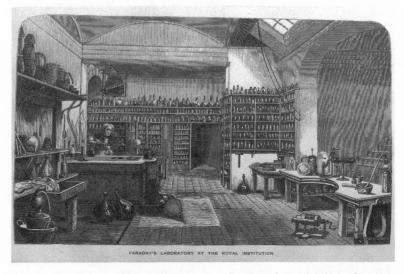

FARADAY'S LABORATORY AT THE ROYAL INSTITUTION.

Figure 7–1. Faraday's laboratory at the Royal Institution. Courtesy Burndy Library, Norwalk, Connecticut.

Eight days later he made an important intuitive leap. Instead of electricity why not use magnets? So he began to explore whether magnetic fields could influence light. He set up the poles of two magnets in five different ways with respect to each other and shone polarized light through glass placed in the field. In four configurations nothing was seen. In the fifth, where the magnetic poles were parallel to each other and light was passed along the field lines, he found that the state of polarization was altered and, moreover, that the effect occurred only with lead glass, not with flint glass, rock crystal, or calcareous spar.

On September 13 he wrote: "Thus magnetic force and light were proved to have relation to each other. This fact will likely prove exceedingly fertile and of great value in the investigation of both conditions of natural force."[3] He was referring to light and matter and could not have foreseen that observations of what is now known as the Faraday Effect would allow twentieth century astronomers to detect magnetic fields between the stars and in distant quasars. In an understatement he ended his diary entry with "Have got enough for the day." It was a great deal more than most scientists would accomplish in their lifetimes.

His faith in the unity of nature, and hence of light, electricity, and magnetism, had been vindicated. He tried to account for the effect of magnetism on light by imagining that the lead glass was in a magnetized state; since electricity produced no such effect it had to be due to magnetism alone. He showed that the glass had no inherent magnetism, so the effect had to be produced by the externally applied field. On September 26 he discovered that the "magnetic force does not act on the ray of light directly (as witness non action in air, etc.), but through the mediation of the special matters."[4] He also found that the phenomenon occurred in many materials. Since the effect was observed only when the glass was placed in the field, it suggested that a field existed within the glass, which then interacted with the externally applied one. In his words, "That which is magnetic in the forces of nature has been affected, and in turn has affected that which is truly magnetic in the force of light." The nature of light, however, remained a mystery. As for magnetism, Faraday now believed that magnetic force had to be present in all things, in all forms of matter. He also knew that some bodies, such as iron, could be magnetized at will and would retain that magnetism. But others seemed impervious to the force. How could he reconcile this with the basic notion that magnetic force was present in all things?

He began by classifying matter into two categories. "Paramagnetics" were substances that were either naturally magnetic, such as iron, or that could easily be magnetized, including other metals such as platinum. Faraday found that these substances were drawn toward regions of stronger field. Lines of force could easily penetrate them. Today it is known that this behavior is the result of the internal molecular structure, which contains what are in essence many small dipoles easily aligned by an external field. The other category, called "diamagnetics," which included salts of various types, acted in the opposite sense. Magnetic fields could not penetrate these substances. Their internal molecular structure is such that they are not naturally magnetic. As a result, diamagnetics could be pushed away by a field.

This categorization caused some confusion about previous notions as to why objects were magnetic. Such theories required the existence of polar molecules that responded to magnetic force just as the magnets themselves responded to the earth's field. But to account for diamagnetism the opposite effect had to be postulated, a reverse polarity as it were. Faraday was skeptical of this approach and was able to show that diamagnetic substances had no poles and seemed to avoid the fields altogether. This is where his notion of lines of force came into play. Magnetic lines of force penetrated paramagnetic substances but could not penetrate diamagnetics. He also showed that lines of force have no beginning or end; that is, they do not originate at one pole and go to another. Instead, they are continuous. This meant that their origin could not be assigned to the existence of molecules, each with their own poles, within the matter that was magnetized.

His great insight was that magnetic lines of force reached out from the magnet. It was as if the surrounding medium acted to complete a magnetic circuit that allowed something to flow in the manner that electricity flows around wires in a circuit connected to a battery. But what flowed?

He wasn't quite there yet, because he still could not explain why magnetism existed, even if he could describe its presence in terms of an invisible field, an aura surrounding the magnet. It had to involve some form of electricity of course. The key point was that "whatever the cause of magnetism, the manifestation of magnetic force took place in the medium surrounding the magnet. This manifestation was the magnetic field and *the energy of the magnetic system was in the field,* not in the magnet."[5] (Italics added.)

That was a key he needed to understand the phenomenon, an insight that would allow other physicists to develop powerful theories to explain magnetic phenomena. The point was that the energy was *in the field,* not in the magnet. The same was true for gravity. The energy of the earth's gravitational field is in the field, not in the ground. What happens to us when we fall in the gravitational field of earth, should we step off the top of a high building, is determined by the field's properties, not by the properties of matter deep in the earth's core. The earth can just as well be made of a pile of old refrigerators, pianos, football stadiums, quarks, molten rock, or water. What it is made of is not the point. The existence and strength of the field were all that theoreticians needed to consider in accounting for gravitational phenomena.

The important conceptual leap was that in order to explain magnetism it was necessary only to deal with the physics of the field, not with the physics of matter giving rise to the fields.

By way of analogy, consider the manner in which water flows down a mountainside. To describe the phenomenon we observe the motion of water along a stream bed. To describe the stream we can draw a map of its course and perhaps even show how it fits into the contours of the surrounding hillside. We can ignore the molecular structure of water and the internal geology of the mountain. Given that we have water we can describe its motion downhill, along river valleys and into the ocean. We do not have to dwell upon atomic theory to understand erosion or the formation of a river delta. Similarly, fields could be used to describe the phenomenon of magnetism, and field theory did not require knowledge about where the fields were rooted. Field theory soon became a science in its own right, with James Clerk Maxwell as the pioneer (Chapter 8).

To account for magnetism in terms of the seat of the force, physicists would have to learn about molecular structure, which was quite unknown in Faraday's time. For example, a bar magnet made of iron has a magnetic field whose pattern is identical to that produced by a current flowing in a loop of wire. Subsequent generations of physicists explored this issue and considered the effect of circular motion of electrons in orbit about atoms, or even of electrons spinning on their own axes, to account for magnetism in solid bodies. While such motion acts to produce fields, it was the *behavior of the fields* that became the key concept that would lead to discoveries that opened the way for the invention of radio and television. We receive (noncable) television in our homes because we live within the "electromagnetic" field (see Chapter 10) created by a distant transmitter. To improve reception we can move the antenna to pick up more energy from the field. We do that without worrying about the construction of the transmitter.

Although Faraday thought he could see "field lines" when iron filings were sprinkled over magnets, those lines were actually figments of his imagination. All he had to do was tap the paper on which the filings rested to see that they moved until they all congregated at the poles of the magnet. Nevertheless, this figment turned out to be a powerful concept for understanding the nature of magnetism. If one wanted to know where the fields originated, well, that was another story, quite unrelated to the physics of fields. In other words, fields should be considered as distinct entities with certain properties, no matter how mysterious they might at first seem. In the mathematical models that were to emerge from these insights, changing the field strength in an equation became a simple matter. The field was a parameter to be manipulated whatever its origin.

But who was to derive the mathematical equations that described the fields whose existence Faraday had intuitively sensed when he studied iron filing patterns? That person was Maxwell, whose ability to put into theoretical terms what Faraday had discovered in the laboratory was the corollary of Faraday's skill at learning in nature's school of experience.

Faraday once wrote to Maxwell as follows:

> There is one thing I would be glad to ask you. When a mathematician engaged in investigating physical actions and results has arrived at his conclusions, may they not be expressed in common language as fully, clearly, and definitely as in mathematical formulae? If so, would it not be a great boon to such as I to express them so?—translating them out of their hieroglyphics, that we also might work upon them by experiment.[6]

This is a perennial problem. To deal with it we have popularizers of science who attempt to translate into everyday language the essence of the truths professionals have discovered in the laboratory, or truths that have surfaced from within an elaborate equation, like steam from a geyser in Yellowstone Park. Unless one captures the essence, the energy is lost to the rest of us.

Faraday believed Maxwell was capable of making the necessary translation so that he, the experimentalist, might understand what the theoretician had seen in his formulas. He went on, "I think it must be so, because I have always found that you convey to me a perfectly clear idea of your conclusions, which, though they may give me no full understanding of the steps in the process, give me the results neither above nor below the truth, and so clear in character that I can think and work from them."

"No man ever felt the tyranny of mathematical symbols more deeply than Faraday."[7] He was always trying to go beyond the limitations of mathematics, trying to transcend its formal expectations. Faraday could understand Maxwell's *conclusions* and was not too concerned with the steps taken to arrive at them. Faraday could sense the way the field and magnet were related. He could *feel* what Maxwell was saying. He could visualize the force independent of any mathematics at his disposal. (It was as if he sensed the "field" produced by Maxwell's thinking and didn't care how Maxwell came to his conclusions.)

For Faraday the magnetic field was "*points* or places characterized only by a certain strength of action." Magnetism was an "interaction of matter with a *property* in its immediate vicinity."[8] It is evident from his drawings of the iron filing patterns that he was impelled to think in new ways about the nature of magnetism as he slowly began to confront the fact that fields were circular. But it is one thing to make the drawings and another to conceptualize what was being sketched. His concept of lines of force was a generalization from his observation of the way iron filings fell around magnets. They literally fell along *lines,* which to him indicated the presence of some invisible force.

The ability to generalize correctly sets the person of creative intellect apart. It surely sets the genius apart from other people. But it is notoriously easy to generalize incorrectly. For example, the pattern of a spiral galaxy may look like the swirls created when milk is slowly stirred into a cup of coffee but that does not mean that galaxies are swirls in cosmic cups. It is one thing to generalize; another to do so correctly.

Faraday's notion of a field was marvelous for its power. It meant that if in field theory the field strength was changed, theoreticians could do that with scribbles on a piece of paper without reference to some unknown physical process acting at the heart of the molecular structure of matter, say. All they had to deal with was the changes in the *field.* This approach to fields as the relevant issue would later allow astronomers to describe magnetic fields in stars or interstellar space without worrying about how the fields are created (see Chapter 14).

Faraday did think that force was a substance, and "that all forces are interconvertible through various motions of lines of force."[9] Maxwell subsequently considered these forces to be stresses and strains in a mechanical ether whose properties were quite unlike ordinary matter. For him the presence of physical lines was a way of allowing the force to pass through space from point to point. These lines could move, were cut by metals placed in their way, and could be bent by the presence of other objects (such as diamagnetic substances). Motions of lines of force might even be able to account for all the forces of nature, he thought. The imaginary lines were quite discrete and connected particles throughout space. They literally came with the magnet, since they connected to its poles.

"We are used to thinking about Faraday as one of the greatest experimentalists who every lived," wrote David Gooding, historian of science.[10] Why was he so good? Because he "was good at learning how to do experiments." The accent is on learning. Faraday's experimental technique was such that he maximized his chance of *learning from nature*. He was an explorer acutely attuned to what nature was teaching him as he probed beyond appearances to learn nature's secrets. As every researcher knows, experiments rarely work first time and in making them work we learn a lot. That was how Faraday learned, by making experiments work where others had failed and given up.

Those of us who are not as purely directed by an ideal in which "laws" are unfolded as the result of experiment may tend to guide our perceptions by unconscious drives that send us in those directions where we *hope* to find the truth. We prefer to look for a lost key in the light of a street lamp because it is brighter there, not because that is where we lost the key. Faraday was not such a person. He searched where nature taught him to search. Paradoxically, he wrote that "there is no [natural] philosophy in my religion."[11] One commentator noted, "Faraday's science, particularly his detailed research, seems so independent of his religious beliefs."[12] Yet he clearly mixed religion and philosophy. To him they were inseparable. For example, he believed that the terms used to describe experimental results should be theory-neutral.[13] "His religious feeling and his philosophy could not be kept apart: there was a habitual overflow of the one onto the other."[14] This directly reflected the Sandemanian credo that he should have no intermediary in reading the word of God. No cleric should stand between him and the words in the Bible. No theoretician's hypothesis should come in the way of experiment if he sought to determine the nature of truth in the physical universe. In a letter to William Whewell,[15] Faraday went so far as to state that promulgating one's "theoretical views under the form of nomenclature, notation or scale actually retard(s)" the progress of science. He was in pursuit of the laws of nature, which his religious upbringing had led him to expect must exist. He was searching for God's laws in nature and became uniquely able to see and recognize the manifestation of those laws more clearly than anyone before him. Although nature's book lies open for all of us to consider, what can be apprehended directly is often subject to distortions introduced by poor vision or is lost in translation.

Faraday was driven by a pure urge, a lofty ideal that remained unfettered by formal university training, which might have forced him into a more conventional mold. He was the right man at the right time doing the right things at the right place. To understand magnetism and electricity he approached the phenomena head-on, without any preconceived notions to misguide and blind him to what might be revealed in his experiments.

Today it would be impossible for a Faraday to emerge from an uneducated background and rise to any significant level in the world of science, given our system of education and the way scientific research is done. The Establishment, our highly formalized educational system, will prevent an "amateur" such as Faraday from entering. Also, the structure of modern science is heavily slanted toward a grounding in theory, which involves learning how to manipulate concepts in mathematical form. I have observed that the system does little to encourage creativity. It does even less to encourage the support of experimentation of the type Faraday was so good at. The modern graduate student in the United States cannot expect to plunge directly into an exploration of nature's secrets without having been thoroughly indoctrinated into how to perform experiments. This is in large part because the equipment that is used nowadays is vastly more complex than anything Faraday could have dreamed of. A modern Faraday would have to approach nature with his or her insight heavily laden with expectations gleaned from lectures. This is generally regarded as a good thing. Viewed from the historical perspective we wonder whether that is necessarily true. Perhaps there is some intermediate way.

Faraday explored nature directly and learned to see more clearly than most others. It is different now. Although science is making huge strides in all disciplines, progress occurs only in the context of expectations that are drilled into students from the start of their studies. As a result, scientists appear surprised when they stumble onto some new phenomenon. But why should they be surprised? I can't help wondering whether Faraday was ever really surprised by what he found. He was surely awe-struck, but was he ever totally taken aback?

NOTES

1. The story of Ampère's missing the boat, as it were, is beautifully reported by James R. Hofmann, "Ampère, Electrodynamics, and Experimental Evidence." *Osiris* 3 (1987): 69.

2. D. K. C. MacDonald, *Faraday, Maxwell and Kelvin.* (New York: Anchor Books, Doubleday, 1964), p. 50.

3. Frank A. J. L. James, "The Optical Mode of Investigation: Light and Matter in Faraday's Natural Philosophy." In *Faraday Rediscovered,* edited by D. Gooding and Frank A. J. L. James. (New York: Stockton Press, 1985), p. 145.

4. Ibid., p. 147.

5. Faraday, *Dictionary of Scientific Biography.* (New York: Charles Scribner's Sons, 1970–1980).

6. Quoted by MacDonald, *Faraday, Maxwell, and Kelvin,* p. 49.

7. Geoffrey N. Cantor, "Reading the Book of Nature: The Relation Between Faraday's Religion and his Science." In *Faraday Rediscovered,* p. 75.

8. Nancy J. Nersessian, "Faraday's Field Concept." In *Faraday Rediscovered,* p. 176.

9. Ibid., p. 183.

10. David Gooding, "Nature's School: Faraday as an Experimentalist." in *Faraday Rediscovered.*

11. Cantor, "Reading the Book," p. 70.

12. Ibid., p. 69.

13. Ibid., p. 75.

14. Ibid., p. 74.

15. Ibid., p. 75.

n 8 n
Maxwell Sees the Light

As I proceeded with the study of Faraday, I perceived that his method of conceiving the phenomena was also a mathematical one, though not exhibited in the conventional form of mathematical symbols.

James Clerk Maxwell,
A Treatise on Electricity and Magnetism

A T about the time Faraday was conjuring up notions about the existence of fields, James Clerk Maxwell (1831–1879) was born in Scotland of a comparatively well-off and highly cultivated family (Fig. 8–1). Before he died at the age of forty-eight he had made marvelous use of Faraday's insights and transformed physics in the process. Maxwell set the study of magnetism and electricity on a solid foundation from which it would never be toppled.

By age three Maxwell exhibited a precocious curiosity by exploring the bell wires that threaded their way through the old mansion in which he grew up. Whenever he wanted to know the answer to some specific question he would ask, "What's the go of that?" perhaps reflecting the notion that if one explored long enough one should be able to find where the trail of mysterious wires led. Should the answer not satisfy him he would insist, "What's the particular go of that?"

Maxwell's nickname at school was "Daftie," a boyish put-down referring to his appearing eccentrically silly or dumb. To those who suffer similar humiliations as a child, there is solace to be found here. Of all the children who attended his school in Scot-

Figure 8.1. James Clerk Maxwell. Courtesy Burndy Library, Norwalk, Connecticut.

land, the Edinburgh Academy, only Maxwell's name will live long in the annals of civilization (although this is little consolation while one is alive!). A biographer later said that Maxwell was not readily fobbed off with dogma, stories, or myths. He maintained a simple Christian faith "that gave him peace too deep to be ruffled by bodily pain or external circumstances."[1]

Perhaps this independence of spirit also made him relatively impervious to the opinions of his youthful peers.

At age fourteen Maxwell published his first scientific paper, on how to draw a perfect oval with a loop of string moving about two fixed points. Halfway through his school courses he underwent a sudden change of personality. A schoolmate reported that after years of shyness, "he surprised his companions by suddenly becoming one of the most brilliant among them, gaining prizes, and sometimes the highest prizes, for scholarship, mathematics, and English verse."[2]

In our book we cannot present a fair summary of a man who studied color vision, discovered the "fish-eye" lens, determined the nature of Saturn's rings, explored statistics and the physics of molecules, and dabbled in engineering. We will attempt only to communicate the essence of his great work, the formulation of a series of equations that summarized all there was to know at the time about electrical and magnetic fields, a formulation that would set the scene for the accidental discovery of radio waves.

Soon after graduating from Cambridge University in 1854, Maxwell began his monumental work on the persistent mysteries of the nature of magnetic and electrical fields by trying to clarify for himself what the problem was about. To proceed he needed to summarize exactly what was known so he could then begin his own journey along the path toward greater understanding. In principle, this is what any good scientist does. In due course his insights would provide an entirely new world to human gaze, a world that lay behind all that Peregrinus, Gilbert, Oersted, Ampère, and Faraday had groped at but could not fully comprehend. Maxwell found and opened a window on the nature of the physical universe that showed rolling scenery stretching to the horizon, scenery of unparalleled beauty and elegant simplicity. If one learned how to step to the window and look out, one could also soak in the view and obtain personal enjoyment from it.

The depth of Maxwell's discoveries, as well as the work of dozens of physicists who came after him, is well beyond the scope of this book. Nevertheless, we can get an inkling of the beauty of the world he found by considering a few highlights. I will try to unveil that vast world beyond our senses, even if we cannot all roam that world beyond appearance, a world that is the essense of the physicist's realm, a world that requires a passport in physics to enter it and explore at will.[3]

When Maxwell arrived on the scene, "he did not have to contend with the morass of uncertainty which Faraday did so much to clear up and the way was more easily open for the key ideas which Maxwell supplied."[4] Faraday had already done the experimental work, and Maxwell, who was to construct the theory, never failed to recognize the value of the contributions made by the gifted amateur of the Royal Institution.

Maxwell believed that the study of electricity was key to understanding nature. This was not obvious to everyone. At the time, scientists were confronting the existence of a bewildering number of forces—gravity, electricity, magnetism, light, heat, and chemical forces. Scientists hoped to sort out what was what by finding common properties among this bewildering array of phenomena. For example, gravity, electricity, and magnetism all showed an inverse square law of force, which Coulomb had determined. That, surely, was significant. But why? What did it mean?

There was something else that disturbed everyone interested in the subject, something we faced when we met Faraday: magnetic force did not act in straight lines. Circles were involved. It was peculiar that when a current ran along a wire and produced a magnetic field, as Oersted discovered, the field did not attract a magnetic needle toward the wire but caused it to orient itself transverse to the wire without drawing it closer. Faraday, as we have seen, accepted this immediately and began to think about the presence of a circular "field." Others, more steeped in tradition found this phenomenon disturbing because it did not fit with what they expected about action at a distance, which required central forces radiating out in straight lines.

A crucial breakthrough, which in some sense made understanding more difficult but which also provided new insights, had been Ampère's discovery that two current-carrying wires attracted one another in a manner similar to the action of parallel magnets. This supported the hypothesis that magnetism had to be electrical in origin. Another concept relevant to Maxwell's work was formulated by William Thomson, who showed that the mathematical equations describing static electricity were of the same form as those that applied to the flow of heat. Maxwell admitted his reliance on Thomson's work very openly. On one occasion he asked Thomson, "Have you patented that notion [of the steady motion of heat] with all its applications? for I intend to borrow it for a season."[5]

He kept on borrowing Thomson's ideas and became self-conscious about it as he jokingly recognized that he was indeed poaching. "I do not know the Game-laws and Patent-laws of science," he wrote to Thomson. "Perhaps the Association may do something to fix them but I certainly intend to poach among your electrical images, and as for the hints you have dropped about "higher" electricity, I intend to take them."[6]

By considering the analogy between heat and electricity, Maxwell was led to the notion that electrical and magnetic forces both involved the flow of something. The term that would be coined for this flow was "flux," which refers to the flowing of a fluid from a body. Magnetic force and electrical force could be thought of as different types of flux. Once the notion of a flow was introduced, the concept of a streamline was not far behind. Maxwell explored this analogy and suggested that Faraday's lines of force were akin to streamlines in a flow pattern. Therefore, what had previously been pictured as almost mystical fields (with an overtone of psychic auras implied) were now to be described in terms of lines of force that began to look rather elegant in their own right.

We cannot follow all of Maxwell's thinking as he developed a theory of electricity and magnetism, except to note that the concepts of fluid, flux, and streamlines lurked in the background. These ideas were combined with the field concept, and the combination became a powerful tool in Maxwell's hands. Fields could be mathematically manipulated, and their description did not depend on the details of the objects (or phenomena) giving rise to the fields. This reminds us that solutions to difficult problems often require a change in point of view. Where curious individuals had once thought that magnets contained effluvia that physically traveled outward in straight lines to act on iron, or that magical influences were due to the soul inherent in lodestone, the explanation for magnetism that finally took hold was the recognition that the magnet was the anchor for a field. The field could then be treated as a thing in itself.

Recall that Ampère discovered that a magnetic field is created by a current flowing through a wire. To deal with the magnetic properties of the field, you no longer had to think about the current. All you needed to concern yourself with was the field, which would continue to exist as long as the current flowed. (Of course, if someone switched off the current the field would cease to exist.) The point was that this approach worked whether a field was produced by a bar magnet or by currents flowing inside the earth. In the case of a common magnet, currents flow on a tiny scale inside the iron. Today we know that they represent the motion of electrons around atoms and of electrons in crystal lattices. But so what? In the case of the earth, the electron flows are believed to be rooted in eddies that swirl within the molten core of the planet. But geophysicists studying phenomena occurring in the terrestrial field where it meets the solar wind high above our planet's surface, for example, do not need to think about what occurs in the core of the planet.

Until this fundamental distinction between the shape and strength of a field and its root cause was fully grasped, no one could begin to understand the nature of magnetism; that is, of that peculiar force that exuded from lodestone and allowed it to influence pieces of iron over a distance.

Now also recall what Oersted, Ampère, and Faraday had found by experiment. A most important key to understanding magnetism and electricity was that these phenomena involved motion, either of electricity in a wire to produce magnetism, or of a magnet with respect to a wire to produce electricity. But as soon as motion was considered, it hinted at the importance of time as an element in any subsequent description. That, in turn, suggested that a finite velocity (a distance traveled in some unit of time; e.g., miles per hour or centimeters per second) was involved. If magnetic effects did occur instantaneously, as was believed by adherents to the action-at-a-distance notion, then neither velocity nor time would be factors in the description.

This awareness turned out to be the foundation of the field equations that Maxwell would discover. The equations describing electric and magnetic phenomena involved time as a parameter, and hence also a velocity. Up to then no one had expected that to understand the nature of lodestone one needed to take time into account.

An electric charge sitting absolutely still does not create a magnetic field. Move the charge, the definition of a current, for example by driving the charge along a wire connected to a voltaic cell, and it produced a magnetic field. This phenomenon underlies the design of an electric motor. Similarly, a fixed magnet did not cause nearby electrons to respond in a continuous manner. But when the magnet was moved, the field cut into the wire and pushed the electrons into motion to manifest as a current flow.

When Maxwell was alive the carrier of electric charge, the electron, had not yet been discovered. Also, he was only interested in a logically consistent mathematical description of these phenomena. His work would explain the nature of magnetism and electricity (especially to those who could read the equations). What he found was that underneath it all, beyond appearances, understanding came from recognizing elegant similarities between the nature of electricity and the nature of magnetism. Above all, the movement of one created the other. Moving charges produced a magnetic field; a moving magnetic field produced a current. At the time few people gave much thought to inventing practical uses for these principles.

It is possible to gain a "feeling" for what is meant by certain concepts that Maxwell used in his theory. The first was *potential*, an elegant notion. A potential refers to something that does not actually exist until it is manifested in an experiment. Yet the concept (label) of potential allowed the calculation of what would happen if an electric or magnetic charge were placed at a given point in space that had a certain potential. The field around a magnet had a potential that was not realized until an electric charge (or another magnet) was placed in that field. Only by interacting with something could the potential of the field be manifested. Until then it had only the *potential* to do something.

The notion of a *vector* as displaying the direction in which the potential acts also comes from Maxwell's work. The vector potential of the magnetic field was described by a direction and a strength of the field at any point in space. These field vectors, as Faraday had found, had a circular shape around current-carrying wires.

The notion of a field of potential allowed for more possibilities than a simple view of action at a distance, which we have discussed before. Action at a distance implied the presence of physical connections between a magnet and a piece of iron, which allowed contact over a distance to be made, like a row of dominoes tumbling down to let one end of the iron rod know about the presence of a magnet at the other end. But how did the field propagate in empty space or through intangible air in the room? The natural explanation was to invoke the existence of an "ether" that filled space so the action could propagate through it. But Maxwell's work would ultimately lead to the rejection of the ether hypothesis.

Another term Maxwell used was *gradient.* Gradient refers to a slope. On a steep gradient on a highway, warning signs tell truck drivers to put their vehicles in a lower gear to prevent a runaway. Physicists use gradient to refer to the rate of change of something over a distance or over time. A rising stock market is an example of a positive gradient in time. The downhill section of a road is a negative gradient in space. In the case of a magnet, the force it exerts on a nail or a paper clip is strong. Farther away the force is weaker, falling off inversely as the square of the distance. It is therefore possible to describe this variation as a gradient (or slope) of the force, how the field strength drops off with distance from the magnet.

The fourth term whose meaning we should "feel" is *divergence.* This concept described the way the shape of the magnetic field varied around a magnet, for example. Close to the pole of a magnet the field lines are tightly crowded and they spread out to be farther apart with increasing distance (see Fig. 7–3). The lines of force describing the field diverge as one moves away from a magnetic pole.

The fifth concept that Maxwell invoked was *circulation.* To produce a mathematical description of a whirlpool, the equations describing the motion of water in swirling eddies, required the concept of circulation. The mathematical description of a skater spinning on ice would also benefit from such a parameter. Circulation, as expressed by Maxwell in mathematical form, thus took into account what Faraday sensed so naturally, that the force of magnetism around a current-carrying wire acted in a circle.

What Maxwell showed was how electricity and magnetism were related using these five concepts, each of which is a measure of some property of the fields, either electric or magnetic. At last the relationship between electric currents and lines of force that Faraday had suspected began to make sense, or at least it did for Maxwell when he discovered the existence of equations that wove vectors, potential, gradient, divergence, and circulation together. These equations would later be simplified by other physicists, whose names were briefly linked to such refinements. In due course four fundamental equations remained and they continue to be associated with Maxwell's name.

Maxwell approached his study, not be considering what others well versed in mathematics had written, but by turning to the words of Michael Faraday, who seemed to be using a language quite different from mathematics. Yet his description had come closer to the nature of reality. This caused Maxwell to write:

> As I proceeded with the study of Faraday, I perceived that his method of conceiving the phenomena was also a mathematical one, though not exhibited in the conventional form of mathematical symbols.[7]

Faraday's tremendous insight awakened Maxwell's imagination:

> Faraday, in his mind's eye, saw lines of force traversing all space where the mathematicians saw centers of force attracting at a distance: Faraday saw a medium where they saw nothing but distance: Faraday sought the seat of the phenomenon in real actions going on in a medium, [the mathematicians] were satisfied that they had found it in a power of action at a distance impressed on the electric fluids.[8]

Faraday's conceptual leap had left others gasping, at least those who were aware of it. It is perfectly natural for us to consider that interactions between objects involve touch, for that is within our daily experience. I cannot move something without physically touching it. As a result I believe that this is natural. Nature, however, does not necessarily function in a "natural" manner. This is where Faraday was able to separate himself from the bias introduced by everyday experience. That was his creative step. It is in this way—by considering a totally new point of view—that problems are often solved.

Maxwell realized that Faraday had begun by seeing the whole, a concept of the phenomenon viewed from a distant perspective. In so doing, Faraday was able to recognize the parts. Mathematical methods, on the other hand, began with the parts and tried to reconstruct a whole. Maxwell was so excited by reading Faraday that he advised all students of science "to read the original memoirs on that subject, for science is always most completely assimilated when it is in its nascent state."[9] How true this still is, and how impossible it has become to follow Maxwell's advice. Unfortunately, to approach the subject with the state of mind of the pioneers is not an efficient way to learn in order to graduate in a university course. Even for the likes of Faraday and Maxwell, it took decades to come to the understanding they struggled so hard to achieve. New generations, if they wish to make progress, cannot afford the luxury of growing at such a "slow" pace by starting all over again. Modern textbooks on electricity and magnetism cannot afford the space to consider the point of view of historical figures. As a result university courses are impoverished through the loss. The great sense of adventure felt by the pioneers of science thus becomes remote.

Faraday was aware of Maxwell's efforts to formalize what he, Faraday, had found in the laboratory. He recognized in the younger man a kindred spirit, even if the two of them appeared to operate in different modes. Many people interested in science have found the mathematics daunting and they will relate to what Faraday wrote to Maxwell in 1857 (we have quoted this before but do so again to make the point):

> When a mathematician engaged in investigating physical actions and results has arrived at his conclusions, may they not be expressed in common language as fully, clearly, and definitely as in mathematical formulae? If so, would it not be a great boon to such as I to express them so?—translating them out of their hieroglyphics, that we also might work upon them by experiment.[10]

Faraday felt that Maxwell had managed to convey something of the spirit of his insights in a way that he could understand. "I have always found that you could convey to me a perfectly clear idea of your conclusions, which, though they may give me no full understanding of the steps in the process, give me results neither above nor below the truth, and so clear in character that I can work from them."[11] He begged other mathematicians to translate their results into a "popular, useful, working state," a cry from the heart to which so many lay people interested in science can still relate.

Maxwell took what he learned from Faraday and tried to generalize the picture. He began with the mental models and experimental results, and after he had found the truth underlying surface appearances he was able to discard the "mental scaffolding."[12] What he was left with was his set of beautiful equations. By 1873 these were available for all to see and use. Not everyone interested and capable of understanding them agreed on their relevance, however. The community of scientists was divided into two politically oriented schools, the British (Maxwell's point of view), and the European, which clung to the notion that electrical and magnetic forces were central and that physical contact through some intervening luminiferous ether instantaneously carried force from point to point (action at a distance).

When Maxwell first looked at Faraday's notion of lines of force he tried to account for them by postulating the existence of a medium that would support the stresses produced by the magnetic lines of force. Wary that analogies can mislead, he thought of a line of force as "a kind of suction tube which drags in fluid ether at one end and expels it at the other."[13] The geometry of the flow along these tubes was supposed to be identical to the pattern of magnetic fields around a magnet. But what sort of medium would support these imaginary tubes?

Maxwell drew upon the work of others in conjuring up the notion that the vibrations consisted of the spinning motion of molecules of gas around the lines of force. He imagined the existence of small "molecular vortices," like so many whirlpools side by side, and aligned so that they rotated at right angles to the magnetic field direction. Because they are spinning they would tend to flatten and thus push against neighboring vortices. To preserve peace and harmony, the vortices would adjust their relative location so they all pushed equally on their neighbors. Thus they would be uniformly spaced. This meant that the magnetic field lines penetrating them would become uniformly spaced. These vortices would produce a form of stress to hold the field lines apart, as is observed around magnets. Close to the poles the field lines would of necessity be forced close together. By making the angular velocity (or spin rate) of each imaginary vortex proportional to the local magnetic intensity, Maxwell found that his formulas were identical to those that existed to account for forces between magnets and steady currents. But when he looked closely at the vortex model, adjacent vortices, where they touched, should rotate in opposite directions. That would not do because then they would cancel out.

What Maxwell realized was that in order to account for magnetic induction—the phenomenon that a changing magnetic field in one place could induce a current in another—he had to allow the change in rotation of one vortex (associated with a change in magnetic intensity) to somehow propagate to another nearby region in space. When one vortex speeded up, how could its neighbor also speed up if at their boundary they were rotating in the opposite directions? He cleverly postulated the existence of little "rollers" between adjacent vortices. Such vortices and rollers did not exist in space, but the notion allowed Maxwell to make progress in his thinking. He could always go back and clean up the mess later, if the thought experiment led anywhere. Then he would present only the mathematical descriptions of the sort to which we have already alluded and leave out sordid details like imaginary vortices with rollers between them.

Maxwell imagined the little rollers to represent electricity disseminated throughout space that was ready to respond to the rotary action introduced by magnetic fields. Notice that the picture is beginning to represent what Faraday accepted so readily, that magnetic fields are circular around a current-carrying wire. Also, this electricity was free to move in conductors but not in insulators.

So Maxwell forged ahead. A speed-up in the rotation of one vortex was coupled to its neighbor through the rollers and almost immediately the whole lot were rotating faster. When the field intensity was changed in one spot, the changing field intensity (as defined by angular velocity of the vortices) was quickly felt farther out in surrounding space. That was how Maxwell pictured magnetism to make itself felt across apparently empty space. The corollary was that the rollers between the vortices formed continuous paths through the medium. The point was that a current could set up rotation of the vortices and so create a magnetic field.

Now imagine a situation that began with no current flowing. Then none of the vortices exist. When a current was sent along a wire and a magnetic field induced in space around it, the field was maintained by the system of vortices. When the current was increased, the field strength changed and traveled through surrounding space. The lines of force behaved as if they were connected and yet they were held apart through the system of vortices and rollers.

Now imagine another wire located parallel to the first. A steady current in the first would drive the vortices, which coupled to the next set, and so on until the movement reached the other conductor. The rollers merely acted to pass on the rotary motion. One could picture them as idling and none of them traveled anywhere. If the current in the first conductor was changed, the vortices speeded up (that is, the magnetic lines of force grew in intensity), and to cope with this the electrical vortices had to change their previously steady state to pass on the change in field (rotary motion). This was felt at the other wire as a current flowing along it.

Consider an analogy of how induction works. If you are standing motionless in a crowded subway train while being pushed up against your neighbors on all sides you will not even notice his or her pressure against your shoulders. Then, if someone near you is jostled you are immediately made aware of it because the impulse passes from one person to the next. In the same sense, if a steady current flows in one wire, the adjacent wire does not notice. Only if the current flow in the first wire changes does the parallel wire notice. When the impulse dies away everything is back to normal again.

Now we come back to Maxwell's picture. Induction involved change, which implied that time had to be brought into the description. That, in turn, implied that a velocity was involved. When Maxwell explored his analogy, he found that he could derive the velocity at which the magnetic impulse traveled through space by using the known physical properties of air around the wires.

The velocity turned out to be equal to that of light, which, by then, had been measured elsewhere. At first his numbers didn't quite match, but as measurements of the velocity of light improved, they were found to agree precisely with Maxwell's calculations.

The stunning conclusion was that electrical and magnetic forces, under the umbrella of a new description, *electromagnetism*, traveled at the speed of light. The corollary was also true, and this too was a wonderful insight. Light had to be an electromagnetic phenomenon! "We can scarcely avoid the inference," Maxwell said, "that light consists in the transverse undulations of the same medium which is the cause of electric and magnetic phenomena."[14]

This was a marvelous breakthrough. After centuries of research, magnetism, electricity, and light were related for the first time. To cut a long story short, and to return to the theme of our book, it is important to stress the *discovery* made by Maxwell. Once he drew back from his vortex model and was able to formally demonstrate the nature of his insights (using his field equations, which involved time as a variable quantity and the speed of light as a constant term[15]), the key to understanding electricity and magnetism was recognized to be the nature of *changing* phenomena. An electric current could produce a magnetic field, and a varying field could induce a current. What was more, these changes were transmitted through space at the speed of light.

In discovering this remarkable truth, Maxwell had set the foundations for the science of electromagnetism, which deals with electricity and magnetism under a unified umbrella. His theory also explained the nature of electromagnetic waves; for example, light. The point was that varying electrical and magnetic fields were together involved in producing light. Light was related to electricity and magnetism, as Faraday and others had long since suspected at an intuitive, Kantian level. It would be several more years before the existence of another form of electromagnetic wave was demonstrated.[16]

Maxwell and others had to abandon the vortex model because of its artificiality, but by then the theory that had emerged from it was firmly established. To repeat our key point, to understand electricity or magnetism it was far more profitable to consider the nature of *fields* and the *theories associated with fields,* rather than the structure of matter that gave rise to the fields. Here Maxwell followed the philosophy proposed by Sir William Hamilton on what is called the relativity of knowledge: "all human knowledge is of relation between objects rather than of objects themselves." It was the relationship between fields produced by "objects" rather than the existence of the objects themselves that was key to Maxwell's work.

We can draw a parallel to the study of human beings. A single individual in isolation does not interact with anyone and is, by definition, unable to learn anything about him or herself. We learn about ourselves only in the context of interaction with other people, and the environment. Psychiatry or psychoanalysis, at their deepest levels, deal with the problems an individual has in relationship to others. It is how we interact with the "fields" generated by friends and acquaintances that may be most relevant to our lives. Or, as a poet once said, "I am all that which I have touched."

Maxwell's discovery of a set of equations that described the nature of magnetism, electricity, and light in one fell swoop, and their successful marriage heralded the discovery of other electromagnetic waves. His equations provided the resolution to the question asked when someone found the first piece of lodestone and noticed that it attracted fragments of iron without touching them. Why?

The answer was ultimately found when the forces of electricity and magnetism were linked into the theory of electromagnetism by Maxwell. The explanation of magnetism required electricity in motion. Inside the structure of lodestone, invisible currents flowed permanently. In turn, a fuller understanding of electricity required knowledge of magnetism. The existence of magnetic and electric fields and their interactions lay at the heart of both issues. But now we leave the realm of everyday experience, because while the notion of a field is still within our grasp, a full understanding of fields requires new tools, those of higher mathematics, and that is beyond the scope of our book. In addition, as we shall see later in Chapter 10, a recent shift in paradigm, or basic outlook, has removed fields from center stage, although field equations still work perfectly well. The deeper explanations involve something even more remarkable.

What is so beautiful about this march of progress is that the original questions about the nature of magnetism led scientists into a new world, a universe filled with electromagnetic waves and electromagnetic phenomena, which could be understood by manipulating mathematical (field) equations based upon experimental discoveries. Even more extraordinary is what we have learned about ourselves in the process of assimilating the new knowledge into our shared archives of data. Through the exploitation of the ideas outlined by Maxwell, and those who followed, we have created a new world and gained precious new insights into our place in the scheme of things on earth and in the rest of the universe.

Electromagnetic phenomena form a cornerstone of our modern world; radio, television, radar, electrical power generation, industrial use of electricity, computers, and the linking of all corners of our planet through ever more sophisticated electronic means depend on our ability to control electromagnetic phenomena. In addition, astronomical studies of the farthest reaches of the universe depend on being able to detect and interpret electromagnetic waves from space. These waves range in length from gamma rays (a billionth of a centimeter long) through light (several hundred thousandths of a centimeter) to radio waves (up to many kilometers in length).[17] These studies, in turn, have revealed that throughout the universe the influence of magnetism is present. And all this because curious individuals once wanted to understand lodestone's magical power of attraction that caused particles of iron to leap up and cling to its surface or why a compass needle could point north–south.

NOTES

1. As reported in an obituary in *Nature* in 1879.

2. Percy Dunsheath, *Giants of Electricity.* (New York: Thomas Y. Crowell Co., 1967), p. 163.

3. Here I drew upon the excellent summary of Maxwell's life and work written by C. W. F. Everitt, *Maxwell* in the *Dictionary of Scientific Biography*. (New York: Charles Scribner's Sons, 1970–1980).

4. R. A. R. Tricker, *The Contributions of Faraday and Maxwell to Electrical Science*. (Oxford: Pergamon Press, 1966), p. 12.

5. Sir Joseph Larmor, *The Origin of Clerk Maxwell's Electric Ideas*. (Cambridge: Cambridge University Press, 1937), p. 11, from a letter written by Maxwell, May 15, 1855.

6. Ibid., from a letter written by Maxwell, Sept. 13, 1855.

7. J. C. Maxwell, *A Treatise on Electricity and Magnetism*. (Oxford, 1904), ix.

8. Ibid.

9. Ibid., p. xi.

10. D. K. C. MacDonald, *Faraday, Maxwell, and Kelvin*. (New York: Anchor Books, Doubleday, 1964), p. 79.

11. Ibid.

12. Ibid., p. 89.

13. Maxwell, *Dictionary of Scientific Biography*.

14. Ibid.

15. The interested reader is referred to any one of dozens of books on electricity and magnetism that present Maxwell's equations in detail.

16. The modern electromagnetic spectrum includes radio waves, infrared or heat, light, ultraviolet radiation, X-rays, and gamma rays, a sequence whose wavelengths run from the very long (many kilometers) to the very short, of the length scale of the nucleus of an atom (less than a billionth of a centimeter).

17. See note 16.

∩ 9 ∩
Heinrich Hertz's
Grand Adventure

With more knowledge comes a deeper, more wonderful mystery, luring one on to penetrate deeper still. Never concerned that the answer may prove disappointing, with pleasure and confidence we turn over each new stone to find unimagined strangeness leading on to more wonderful questions and mysteries—certainly a grand adventure.

Richard Feynman,
What Do You Care What Other People Think?

Around the turn of the century commercial exploitation of the facts unearthed as a result of pure research was about to begin on a vast scale. Electricity was becoming widely available and its practical application was imminent. Arguments began to rage as to whether to use direct current or alternating current to supply homes and industry,[1] although most physicists gave little thought to the commercial implications of their work. They continued to search for more comprehensive explanations for both electricity and magnetism, in particular within the context of the new science of electromagnetism.

Concerning the social implications of the study of magnetism, Maxwell wrote, "It is hardly necessary to enlarge upon the beneficial results of magnetic research on navigation."[2] But he underestimated the future impact of such studies when he wrote, "The important applications of electromagnetism to telegraphy have also reacted on pure science by giving commercial value to accurate electrical measurements," and suspected that this work would likely contribute to the "general scientific progress of the whole engineering profession." It did far more than that. It would totally alter life for virtually everyone on earth.

Thus it was that in the late nineteenth century several more players appeared on the stage to fulfill their roles in clarifying for themselves what Maxwell had found in his equations. Their work broadened the quest for answers to basic questions and create the setting from within which electromagnetic knowledge would be used to develop radio, television, and radar.

One of the players we will meet, Heinrich Hertz, can be said to have single-handedly changed the course of human history. But first we must visit Hermann von Helmholtz (1821–1894), the great German physicist who, at the University of Berlin, began to play a pivotal role in making Maxwell's work accessible to the European school.

Helmholtz's family could not initially afford to send their son to university unless he chose to enter medicine for which state financial aid was available. He obtained his M.D. in 1842, but before long his wide-ranging interests turned to science. His prodigious genius was such that one biographer wrote that he was "the last scholar whose work, in the tradition of Leibniz, embraced all the sciences, as well as philosophy and the fine arts."[3] It is virtually certain that no one human being can ever do so again, owing to the growth in the last hundred years of scientific knowledge in the areas of the physical, chemical, and biological sciences.

After doing research in physiology, Helmholtz turned to the study of conservation of energy, an interest driven by his reading of Kant. As was true for so many who had been inspired by Kant before him, Helmholtz was convinced that two abstractions, matter and force, had to underlie the nature of phenomena and by seeking hard enough one would find the unifying principles that held all things together. Experimentation should make it possible to trace the ultimate cause of all phenomena, because in Helmholtz's time everything was believed to be the consequence of cause and effect.

It is Helmholtz's exploration of electrodynamics that concerns us. The year was 1870 and Maxwell's work was barely known in Europe when Helmholtz turned his attention to the subject. He found a "pathless wilderness" of competing mathe-matical formulas and theories that had been proposed during the past fifty years.[4] At the time the majority of physicists were constantly refining and redeveloping variations on themes set forth by others. To Helmholtz little progress was apparent and he wanted to demonstrate experimentally which of the compet-ing theories for electromagnetic phenomena was closest to the truth.

According to Maxwell, electrical disturbances traveled through a dielectric medium (in which constant currents cannot flow but in which varying currents do propagate) at the velocity of light. Helmholtz developed more specific aspects of Maxwell's equa-tions and drew the attention of other physicists in Europe to the Scotsman's work. In principle, the two theories popular at the time, instantaneous action at a distance versus the Faraday/Maxwell field concept, should have made predictions about the effects produced by electricity in motion (a field of study called electrodynamics) that contradicted one another. Helmholtz ap-preciated that a key to determining which theory was correct would involve unclosed currents; that is, currents that flowed back and forth in circuits that were not physically closed. An example of such a circuit is a loop of wire terminating in a spark gap. A steady current could not flow unless a circuit was complete. When a spark jumped across the gap, the circuit was momentarily complete, but then what? It was also not obvious what would happen to changing or oscillating currents in the same situation. How would they react to the presence of the gap?

Picture the nature of a television or FM antenna of the type that is either mounted indoors (e.g., a set of rabbit ears) or out-doors on the roof of a building. Such antennas, known as di-poles, consist of two pieces of metal connected to wires, which the hardware store clerk will tell you is called twinlead. Seen from the point of view of the electrical circuits involved, the antenna circuit is open, terminating as it does at two metal rods (or rabbit ears). It was the exploration of an analog of this type of circuit that was considered by Helmholtz.

There were two other phenomena that interested him as well: induction, which involved the movement of electricity through space; and ponderomotive effects, that is, movement

of the conductor as a whole under the influence of electrodynamic forces. Ampère and Faraday had discovered such mechanical effects when current-carrying conductors caused movement
in nearby conductors. Helmholtz expected that ponderomotive
effects should result from the piling up of free electricity at the
ends of open circuits.[5] However, his initial experiments with a
number of collaborators did little to help clarify the confusion
of theories, and in 1876 his interest waned, although in 1881
he predicted the decline of the action-at-a-distance theory and
believed that Maxwell would be shown to be correct.

The seed sown by Helmholtz when he realized that the crucial test to decide between action at a distance and Maxwell's
view of electromagnetism would involve open circuits fell onto
fertile ground, even though the seed took many more years to
germinate.

At this point Heinrich Hertz (1857–1894) entered the scene
(Fig. 9–1). Here was another of our characters who entered
physics having started in a totally different profession, in his
case architecture. It is almost an anomaly in the history of science that Hertz was born into a wealthy family. By the age of
eighteen he was a student of engineering but was developing a
side interest in mathematics and physics. This interest grew as
he read about the great historical figures in the area of electricity and magnetism, readings that caused him to lament in a
letter to his parents:

> Sometimes I really regret that I did not live in those times, when
> there was still so much that was new; to be sure enough [much]
> is yet unknown, but I do not think that it will be possible to
> discover anything easily nowadays that would lead us to revise
> our entire outlook as radically as was possible in the days when
> telescopes and microscopes were still new.[6]

"Although this feeling may be to some extent wrong," he confessed, concerning his lack of optimism, "I do not believe that
it is completely unjustified."[7]

This sad note of whimsy was expressed while he was awash
with new information gleaned from books in which he saw the
beauty of science contrasted with the absurdity of some of men's
endeavors, such as the work of a Jesuit who asked which would

Figure 9–1. Heinrich Hertz. Courtesy Photo Archive, University of Karlsruhe, Germany.

be a greater miracle, "a gnat one yard long or a whale 1000 yards long?" or whether Eve should be regarded as the daughter or sister of Adam.

Hertz was on a career track that would make him an architect, although, as with so many young people of eighteen, he was far from ready to make any career decision. In 1901 his mother, Anna Elisabeth Hertz, looked back on those years in Heinrich's youth with fondness and, in terms so many parents who value letters from their children in far off places can relate to, she wrote that after he left home "it was his letters, written regularly and giving full accounts, that made us happy."[8] She ends her letter with a quote from one of Hertz's:

> I intend, if I succeed in passing the matriculation examination, to go to Frankfurt am Main and work a year under a Prussian architect, as I would ultimately be required to do for the state licensing examination for professional engineers. Only if I were proved unsuited for this profession or if my interest in the natural sciences were to increase further, would I devote myself to pure science. May God grant that I choose whatever I am best suited for.[9]

In April, 1875, Hertz began to work at the public works office in Frankfurt where he soaked up the writings of the classical scholars of Greek and took up sculpture in his spare time. He soon tired of that existence, which seemed fruitless. He managed to get a job with an architect but in the meantime had heard about a physics club in town. To satisfy his craving for more knowledge, he attended a meeting and at this point became even more fascinated with science. When he was warned by colleagues that "a Prussian state-licensed architect not only has poor pecuniary prospects but is also looked upon as a machine working for the Prussian state and generally comes to nothing,"[10] his future seemed bleak indeed. "The last part bothered me very much," he wrote. "I will not believe it."

His inner conflict continued. Similar struggles to decide between a career thrust upon you by either family or society and one in which you are free to pursue your deepest interests have wracked many a person before and since. In this case the moral of the story is that to live a full life one must, ultimately, give in to what feels best. As Henry Miller once wrote, "Every man has his own destiny: the only imperative is to follow it, to accept it, wherever it may lead."[11] Tuning into what destiny has in store can be very difficult, however. For too many individuals the luxury of making a choice might not even seem to exist, although the experience of Faraday and Hertz offers encouragement to young readers who might be struggling with a similar dilemma.

Hertz's struggle raged for many years, as was revealed in his diary entries where he confronted the options open to him. Should he pursue a career as a respected architect or give in to his deeper instincts and become a scientist? He fought to suppress his feelings. "I do not let my old desire for studying the natural sciences rise uppermost again." [12] But he continued to read more about physics in a book he found in a local library, [13] and which set him to performing his own experiments. He built his own test equipment, some of which, including a galvanometer, he would later use as a graduate student.

One day, "Before lunch I went to an optician to find the price of a battery and an electromagnet; but he said he did not stock the inferior sort and those he had were too expensive for me." [14] So the fledgling architect tried to obtain permission to use the local Physical Society's resources. He wrote to its president, who replied that he would take the request to the board. A month or so later, Hertz withdrew his request because he felt he had been presumptuous even to imagine he could use their facilities. However, he did have an idea concerning telegraphy that he wished to test, but his hesitation to use the laboratory was reinforced when he read a book on the subject. "[I] discovered that my idea was the fundamental concept of the entire field of automatic telegraphy." The system he had imagined was already obsolete, and he felt embarrassed that he had dreamed he could achieve something new in the world of science.

Hertz continued to read voraciously and soaked up books on chemistry, heat, and even economics. Then he began to experiment with light. Finally he made the break, left the architectural route and decided to study mathematics. In April, 1876, he went to Dresden to attend the Polytechnic where his craving for science could be satisfied. However, this step did little to enhance his sense of self-worth.

Day by day I grow more aware of how useless I remain in this world. I know a little Greek and a little mathematics, a little this, and a little that, and now I have learned to play billiards. [15]

But within ten years his worth would be proven beyond measure. His letter continued:

> I have greater expectations of the future than satisfaction with the past. I believe I wrote you in a very similar vein a year ago, and really the entire difference between now and then is that I have become a year older, otherwise I have made no progress.[16]

August, 1878, found him in Hamburg where he performed experiments with a small but very sensitive compass. He was astonished at how sensitive the compass was to a small magnet placed at some distance and suspended in water while completely enclosed in glass and wood. He was becoming aware of the phenomenon of action at a distance.

On October 21, 1878, Hertz went to Berlin where introductions from former professors landed him a place at the university. There he attended lectures by Helmholtz, who advised him on what reading to do to become more involved. He encouraged Hertz to tackle a prize question that Hertz believed fell into an area that already interested him.[17] A prize question was a research topic that one pursued in competition with other students. The best solution earned the prize. This particular question concerned whether electricity moved with inertia; that is, whether it would continue to move after the circuit was broken, or whether it would take time to begin moving when the circuit was closed. In other words, how long did electricity take to react when a current stopped or started? The task was to determine whether an extra current could be detected over and above that which one would intuitively expect upon switching on or off. He was unsure of the outcome of his research when he wrote:

> I must also mention that, for the time being, I am only tentatively working on it, and since I may fail I want no talk of my work being for a prize problem.[18]

Hertz was deeply wracked by the uncertainty that is a hallmark of the creative mind.[19] He did not want to be seen as having failed, and only mentioned his work to his parents by way of explaining what he was doing with his time. He felt he had to justify his existence to his father who was sending him money to live.

To measure currents in his experiments he used his home-made galvanometer, which he mounted on a wall. It shook with the passing of horse-drawn wagons and that made measurement very difficult. He built his own batteries in an adjacent room and sent the current through wires that led through the wall, which kept him away from the poisonous battery fumes. He wrote that "Helmholtz comes in every day for a few minutes, looks at what is going on, and is very friendly."[20] Meanwhile Hertz was beginning to suspect that electricity had no inertia and confessed that a negative result was less satisfying than a positive one. He also felt a little guilty because he had not signed up for as many courses as other students and spent so much time in the laboratory on his research.

> I cannot tell you how much more satisfaction it gives me to gain knowledge for myself and for others directly from nature, rather than merely learning from others and for myself alone.[21]

Hertz's feeling of wanting to make an original contribution was so strong that he admitted that "As long as I work only from books I cannot rid myself of my feeling that I am a wholly useless member of society."[22] His knowledge of electricity was still rudimentary, but he marveled at the fact that six months before he had hardly understood anything of the subject. This also worried him lest it impede his progress.

Helmholtz recognized the talent of the young man who had dedicated himself to becoming a "great investigator." On August 3, 1879, Hertz was informed that he had won the prize (electricity had no inertia) and received additional praise for a job well done. This caused him to respond modestly, "the evaluation I received was 5 times better than the work merited at best in my own opinion, and 10 times better than I had expected."[23] He wrote that when he went to attend the judgment he was prepared to let nothing show if the result had been unfavorable. The joy he actually felt was so great that he later claimed to feel ashamed over his reaction because he considered the accomplishment so small when seen in a larger context.

He then turned down the opportunity to tackle another prize problem, also set by Helmholtz, which was meant to decide the issue that intrigued the professor. This was related to Helmholtz's question about which of the competing theories for electromagnetic (or electrodynamic) phenomena was correct? Was it action at a distance or the field model of Faraday and Maxwell? Helmholtz believed that a decision depended on making suitable measurements involving open circuits, and the prize was offered to the experimenter (explorer) who found a way to make a judgment.

Hertz realized that it would take him at least two or three years to come up with something satisfactory, and because this would again be competitive work he did not like the idea of doing it in secret. He refused the challenge.[24] Instead he decided to do an experiment on the induction produced by rotating spheres, a project he had started at home and had wanted to finish for some time. His work topic progressed so fast that he was able to graduate with honors well ahead of schedule (after four semesters). Despite his obvious successes he was relieved that he had avoided the "ridicule of failure"[25] in going for the prize question.

Although Hertz took no action to work on Helmholtz's pet topic, the question it posed sank into his unconscious where it lay dormant for seven years.

In 1883 Hertz found himself entrenched at the University of Kiel where he was frustrated by a lack of equipment or funds to purchase any new items. His diary contained terse entries that reflected that not much had happened to him since he solved the problem of the inertia of electricity back in his Berlin days (nearly five years before). His latest crisis of self-doubt reached a peak when one day none of his students came to class.[26] However, he did think about Maxwellian electromagnetism and wrote a paper on Maxwell's equations but "for the most part thought over electromagnetics fruitlessly."[27]

Dissatisfaction at Kiel led to a job change. Even then he was hardly enthused about his new position, professor of physics at the Technische Hochschule at Karlsruhe, where he arrived in late March, 1885. Soon after he arrived there, a dark mood brought about by loneliness descended over him. By then Hertz had become an ardent reader of the works of—guess who?—

Immanuel Kant. He came under the influence of the philosopher's ideas related to the intrinsic unity of forces in the universe. This did little to relieve his persistent depression:

> I am well, but for the time being I still have a great deal to do.
> Or rather I have nothing to do than what I shall have to keep
> on doing for the rest of my life, in all probability; but every be-
> ginning is difficult, and for the time being each day I dread the
> next.[28]

He sank deeper into the slough of despair as he lived and
ate alone in his apartment:

> The first day I thought I could bear it only temporarily, but now
> I have been eating alone for three weeks and could continue for
> a year. For no longer than that, I hope; for if I am not married
> by the end of the year I shall be in a boundless fury.[29]

He began to potter about with experiments at the college.
At this point no one would have predicted that great things
were on the horizon. He felt himself to be just another fairly
bright, young physics professor at a small college in Germany.

One fateful day he found several induction coils in an old
equipment cabinet. A flicker of enthusiasm stirred within him.
Perhaps he could do some interesting experiments with those
coils. Since he did not suffer from precognition he had no idea
that these coils would help make him world-famous. His dark
mood persisted. "I have already given up hope of finding pleasant company of my own age, such as there was at Kiel,"[30] he
wrote. Hertz was really lonely, and with a difference. Previously
he had been able to work while lonely. Now he couldn't even
work because there was nothing obvious to do.[31] He began to
resign himself to being a teacher and very little else. This led
to his lament that "Will I be one of those who cease to contribute anything after they achieve a professorship?"[32] (Today, an
analogous feeling has been known to overcome those who have
been given tenure at a university.)

He began teaching a course on meteorology for farmers and
"tortured myself again with futile thoughts,"[33] which were in-
terspersed with solitary walks through snowstorms. He "strug-

gled hard with ill-humor and lack of hope throughout."[34] At the end of 1885 he told his diary that he was happy the year was over and that he did not want another one like it.

In January Hertz had a bad cold and a toothache, the first symptom of an illness from which he would never recover. Then, quite suddenly, his diary announced that he was engaged to be married to Elisabeth Doll.[35] On July 31, 1886, a single entry stated that it was their wedding day. His loneliness was over, or so we must surmise. His sister wrote later that "His first great scientific successes began to materialize at the same time almost *as if by magic*."[36] (Italics added.) This is reminiscent of the transformation that was said to have overcome Maxwell in his school days when he suddenly made a great leap forward in his intellectual life. (Similar transitions have been noted in the lives of many significant historical figures.[37])

And so it was that the scene was set to change the course of history. Heinrich Hertz began to experiment with sparks in a short metal loop attached to an induction coil. By connecting its primary winding to a source of electricity, such as a voltaic pile or Leyden jar, it was possible to induce far greater voltage in the secondary winding. If the secondary winding was left open and allowed to terminate at two metal balls separated by a small distance in air, impressive sparks could be generated.

In setting up his apparatus, Hertz used the induction coils he had found in the old closet. Such coils were common in late nineteenth century laboratories, but he used them with a difference. To be able to vary the electrical characteristics of the circuit, he connected two metal rods with spheres on their ends to the two sides of the spark gap (Figs. 9–2 and 9–3). We can recognize this configuration as a simple radio antenna in the making, but of course Hertz didn't have a clue about either antennas or radio; no one did.

On November 1, 1886, Elisabeth was with him in the laboratory when the momentous event occurred. He caused sparks to jump across the gap and was startled to observe small "side sparks" where they had no right to be. Some historians have estimated that they may have been only one-hundredth of a millimeter long.[38] These miniscule electrical discharges, triggered in some mysterious way by the large spark in his apparatus, would ultimately lead to the invention of television and

Figure 9–2. Hertz's apparatus, which he used to explore the nature of the "electric waves" generated when the induction coil at the left sent a spark between the two small spheres at the center. Large spheres were connected by means of metal rods to the spark gap. Those rods acted as a radio antenna, which transmitted the energy that produced the un-expected "side sparks" observed by Hertz. Courtesy Photo Archive, University of Karlsruhe, Germany.

provide astronomers with the key that allowed them to bridge the universe (see Chapter 13).

Hertz could easily have overlooked this peculiar phenomenon. Keen eyesight and pitch darkness were required to see the side sparks and they certainly were not expected. He did not report exactly where they were seen, nor whether it was he or his wife who first noticed them. But it was he who began a relentless quest to understand what was happening. He responded to this accidental discovery[39] with his curiosity stimulated and systematically began to explore the nature of the mystery spark.

In the realm of nature, accidents often define the path of biological evolution, as has been so convincingly argued by Stephen Jay Gould[40] and David Raup.[41] But of course we can never

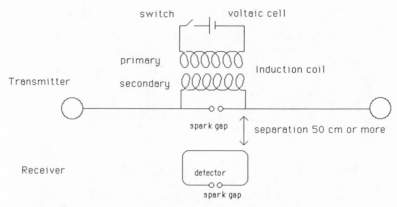

Figure 9–3. A diagram of Hertz's apparatus shown in Figure 9–2. The induction coil vastly amplified the voltage applied from a voltaic cell connected to the primary. The spark jumped across the gap in the secondary circuit. The metal rods acted as a radio antenna. The detector Hertz used was another spark gap connected to a small metal coil.

prove how things would have been had the accidents not unfolded the way they did. All we can discern is how the present has been determined by such events. In other words, we examine history to explain how things turned out to be the way they are, but we cannot replay the tape of technological progress, or of life itself, to see how things would have been otherwise if such-and-such had not happened. We can only suggest that if certain major steps had not been taken, things would be different today. This is particularly so in the case of Heinrich Hertz. First it turned out to be a quirk of fate that caused a side spark. A suitable piece of equipment unrelated to his main apparatus had to be in just the right place or the spark would not have happened. Second, he or his wife had to notice it. Historians of science later discovered that a similar effect had been seen by a half dozen others before Hertz, but they all ignored the message. Also, Hertz had to decide to proceed with his experiments. Had any one of these factors been otherwise, our lives would be different today, particularly in those parts of the world that rely on radio and television.

So there he was: after years of frustrated ambition and lack of direction, Hertz was suddenly involved in a grand adventure.

He was confronted by a clear-cut mystery that had to be solved. Like so many other explorers before him, he had no idea of where he was headed. But he did have the intuitive skill to move in the right direction, step-by-step, as he unraveled the factors that controlled the nature of the side sparks.

His first step, of sending the current from an induction coil to a spark gap, was not unique. Many others had done that before.[42] In summarizing his own work, Hertz later admitted that "it would have been impossible to arrive at a knowledge of these phenomena by the aid of theory alone."[43] Concerning the unexpected sparks, "Their appearance upon the scene of our experiments depends not only upon their theoretical possibility [that they are allowed in nature], but also upon a special and surprising property of the electric spark which could not be foreseen by theory." Since he was familiar with Maxwell's equations, and had already done work to modify them, this statement leaves no doubt that he was initially in the dark about the meaning of his discovery. But Hertz's description was not quite correct. It was not the property of the side sparks that was important. It was the fact that the primary spark was oscillatory that turned out to be crucial. The crackle of the primary spark involved rapid changes of the electric current that jumped across the gap during the discharge. It was possible to measure the frequency of this oscillation by looking at the spark in a rapidly rotating mirror, which acted like a stroboscope.

Hertz found that he could produce sparks in a secondary circuit at will, provided it was placed relatively near a primary circuit that was sparking. In this way he learned to bring the phenomenon of the side sparks under control. Only then could he begin to experiment to find what factors determined their presence and their characteristics. For this he built a detector that consisted of a small metal loop that terminated in a spark gap whose spacing could be accurately adjusted with a micrometer. With this remarkably simple and, from our standpoint, crude detector he determined where the sparks were the longest and how their length varied with position. That measurement told him how much energy was present at any given location to produce the side sparks. When he drew a picture (a graph) that showed length of spark at different positions, it produced a wave pattern running from zero to a maximum and then back to zero again. This meant that an invisible wave, an "electric" wave as he called it, was present in his laboratory. His detector was used to map out the shape of the wave, from zero to low to high and back to low and zero again, as he moved the detector away from the primary spark.

The distance between the locations where the spark was nonexistent gave the length of the wave. Using the rotating mirror, Hertz determined the frequency of the wave from the spark discharge and then combined frequency and wavelength in a simple formula to determine velocity.[44] Now he knew how fast the wave was traveling.

The "electric" waves were traveling at the speed of light! But they were not electric waves. From his systematic exploration, Hertz realized that he had found a new form of electromagnetic radiation. Something like light, but with a much longer wavelength,[45] these invisible waves existed within his laboratory. They were radio waves. Quite unknowingly, he had stumbled onto a major discovery.

Hertz's discoveries formed the core of his most important work. "The finding out and unravelling of these extremely orderly phenomena gave me peculiar pleasure,"[46] he admitted, no doubt one of the great understatements by any scientist in history.

He began to explore analogies between light and his newly identified "electric" (radio) waves and showed that they could be refracted (bent) in passing through materials such as pitch, diffracted (broken apart) in passing through narrow holes in metal screens, and could be polarized (see Chapter 12 for a definition of polarization) by passing them through a grid of parallel wires. He even learned to focus the "electric" waves using large concave mirrors (Fig. 9–4), the forerunners of the metal dishes now so common in satellite communication and radio astronomy.[47]

Contrary to many popular expositions of Hertz's discovery, he had no prior expectation as to the existence of radio waves. Many authors have suggested that Maxwell predicted the existence of such waves, but that supposition can only be made with imaginative hindsight. There is no evidence that either Maxwell, Helmholtz, or Hertz predicted the existence of radio waves.

One historian who devoted a lengthy article to Hertz wrote,

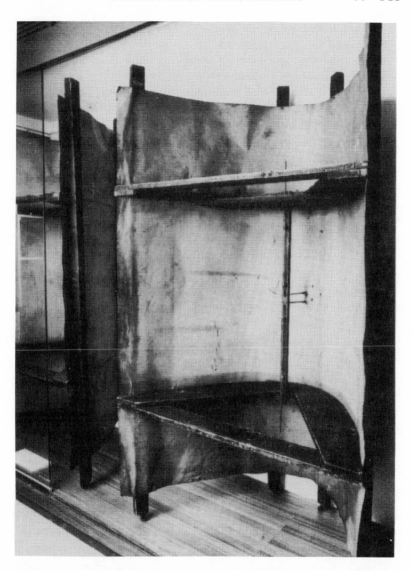

Figure 9–4. The reflectors Hertz used to prove that the "electric waves" could be focused and which he used to transmit these waves across the full width of his laboratory. Courtesy Photo Archive, University of Karlsruhe, Germany.

The whole first volume of Maxwell's Treatise deals with electro-
statics and steady currents, and only a limited number of pages
of the second volume deal with the theory of the electromagnetic
field. In the second volume no trace can be found of a theory of
electric oscillations. More important, perhaps, is the absence in
Maxwell's writings of any theory connecting a propagating field
and an oscillating current as its source.[48]

But it was the oscillating current that gave rise to the radio waves
and Hertz's discovery was totally unexpected.

It is very possible that Hertz's observation of the unexpected
side spark may have been one of the greatest serendipitous discoveries in the history of science. Others had overlooked a similar phenomenon,[49] and initially Hertz was quite reluctant to
pursue his experiments because, as he wrote later, "At first I
thought the electrical disturbances would be too turbulent and
irregular to be of any further use."[50] But then, using his spark
detector, he discovered the existence of neutral points along the
length of the secondary circuit where no spark could be seen
with his detector. This finding was a glimmer of order in what
at first had seemed chaotic, with sparks appearing where they
had no right to be. Now he had something to guide further
experimentation.

In his early descriptions of his work, it is clear that he thought
he was dealing with induction, not with electromagnetic radiation in the form of radio waves as we now understand them.
He was exploring the nature of what induction was, and on
November 13, 1886, he explicitly stated that he had "succeeded
in demonstrating induction between two open circuits; current
circuits 3 m long, distance apart 1.5 m!"[51] He had transmitted
radio waves over the enormous distance of 1.5 meters (about 5
feet).

It took months for Hertz to nail down the properties of this
enigmatic phenomenon. When he was satisfied that he knew
enough about the mystery, he wrote a letter to his former teacher,
Hermann von Helmholtz in Berlin. As a cautious scientist confronting something very significant, Hertz admitted that "there
is an obvious danger of errors and false interpretation in all
these experiments, but I have already found my conjectures

confirmed in a sufficiently large number of experiments to convince me firmly that my interpretation of the phenomenon is substantially correct."[52]

That is more than can be said about his opinion of his medical problem. "For my own person I have some cause for complaint," he wrote to his parents on January 15, 1888. "So far this has been a year of toothaches and dental operations, and at the moment my right cheek is as swollen as my left one was earlier, and I cannot bite on a single tooth. Thank God one has learned in a long life to recognize this sort of condition as transitory."[53]

But it wasn't transitory; six years later, at the age of thirty-six, he was dead. Hertz was a scientist experimenting at the frontiers of the known world and in that setting he would not have dreamed of resorting to wishful thinking or superstition to account for his observations of natural phenomena. As to his perpetual medical predicament, what could he do except resort to hope? After all, no one knew how to cure the tooth infection.

He continued as best he could to concentrate on his laboratory work. It turned out that the original side spark had manifested itself only because an appropriate conductor in his apparatus was located at the position of an energy maximum radiating from the primary spark. We can only wonder how his research, and the history of science and technology, would have been different if this fortuitous circumstance had not occurred. Also, the sparks he studied were often so weak that it is hard to imagine that he would ever have searched for them without already having an awareness that spurious side sparks existed. At least six prominent physicists were subsequently found to have seen the same effect before Hertz, but each of them dismissed the phenomenon as not worthy of further study.[54] How much longer would the world have waited for the discovery of radio waves if Hertz had not stumbled onto his side sparks and then given them his undivided attention?

Hertz's work had also, inadvertently, led to the solution of the dilemma that had so troubled Helmholtz. The discovery of radio waves, and the way in which an oscillating spark could produce radio waves traveling at the speed of light, came down firmly on the side of Maxwell. The notion of instantaneous action at a distance ceased to be of much use. This was in 1887, four decades after Maxwell began to uncover the foundations of his theory and nearly two decades after Maxwell's work was widely published.

Tragically, Hertz never lived to appreciate the magnitude of his discovery. On January 1, 1894, he died of blood poisoning associated with his apparently trivial toothache.

> Hertz entered physics at the right time for one of his abilities to make a critical contribution; because the outstanding problem in physics was the disorderly condition of electrodynamics, what was needed was someone with the theoretical power to analyze competing theories and with the experimental judgement to produce the evidence that would persuade the physical community that a decision between the theories had been reached.[55]

Had he lived into his seventies, which was the lot of so many of the physicists who experimented with magnetism and electricity, he would have seen how radio waves would allow communication between the farthest corners of our world. Had he lived he would undoubtedly have been the first winner of the Nobel Prize in physics, an award that was not presented until 1899.

Aware that he was not recovering from his dental problems, Hertz stoically wrote to his parents a year before his death:

> If anything should really befall me, you are not to mourn; rather you must be proud a little and consider that I am among the especially elect destined to live for a short while and yet to live enough. I did not desire to choose this fate, but since it has overtaken me, I must be content; and if the choice had been left to me, perhaps I should have chosen it myself.[56]

NOTES

1. For a fascinating account of this controversy, and the role played by one of the great pioneers in the study of electricity, Nikola Tesla,

the reader is referred to *Tesla: Man Out of Time* by Margaret Cheney (New York: Dorset Press, 1989).

2. J. C. Maxwell, *A Treatise on Electricity and Magnetism.* (Oxford, 1904), p. vii.

3. Helmholtz, *Dictionary of Scientific Biography.* (New York: Charles Scribner's Sons, 1970–1980).

4. Ibid.

5. Ibid.

6. *Heinrich Hertz: Memoirs, Letters, Diaries.* Arranged by Johanna Hertz. (San Francisco: San Francisco Press, 1977), p. 81. Hereafter referred to as *Heinrich Hertz.*

7. Ibid.

8. Ibid., p. 21.

9. Ibid., p. 27.

10. Ibid., p. 37.

11. Henry Miller, *Reflections on Writing.* Quoted in *Henry Miller on Writing,* edited by Thomas Moore. (New York: New Directions, 1964).

12. *Heinrich Hertz,* p. 39.

13. Wüllner's *Physics.*

14. *Heinrich Hertz,* p. 39.

15. Ibid., p. 55.

16. Ibid.

17. Ibid., p. 93.

18. Ibid., p. 95.

19. Rollo May, *The Courage to Create.* (New York: Bantam Books, 1975), p. 13.

20. *Heinrich Hertz,* p. 99.

21. Ibid.

22. Ibid.

23. Ibid., p. 113.

24. Ibid., p. 115.

25. Ibid., p. 119.

26. Ibid.

27. Ibid.

28. Ibid., p. 205.

29. Ibid.

30. Ibid., p. 209.

31. Ibid.

32. Ibid.

33. Ibid., p. 211.

34. Ibid.

35. Ibid.

36. Ibid. From preface to second edition, by Mathilda Hertz, p. xii.

37. As described in 1901 by R. M. Bucke, *Cosmic Consciousness.* (Reprinted by Citadel Press, New Jersey, 1977).

38. Joseph F. Mulligan, "Heinrich Hertz and the Development of Physics." *Physics Today* (March 1989):52.

39. An event that happens by chance or has unforeseen causes.

40. Stephen Jay Gould, *Wonderful Life.* (New York: W. W. Norton, 1989).

41. David Raup, *Extinction: Bad Genes or Bad Luck?* (New York: W. W. Norton, 1991).

42. I spent many hours playing with an induction coil as a teenager and caused sparks to leap all over the place using common batteries to feed an impulse into the primary. Alas, I was already so accustomed to the existence of radio waves that when I heard the spark create a sound on the loudspeaker of a nearby radio it seemed perfectly natural. For the pioneer Hertz, however, there was no radio to detect the fact that the spark created a radio signal that traveled through the air. In fact, his method of demonstrating that something mysterious was generated by the spark, and that it crossed considerable distances, was as inspired a piece of work as was ever witnessed in the history of physics.

43. Heinrich Hertz, *Electric Waves.* (New York: Dover, 1962), p. 3.

44. Velocity = frequency times wavelength ($c = f\lambda$).

45. The waves Hertz detected were meters long, unlike light waves, which are around 10^{-5} cm in length.

46. Hertz, *Electric Waves,* p. 5.

47. Gerrit L. Verschuur, *The Invisible Universe Revealed.* (New York: Springer-Verlag, 1987).

48. Salvo D'Agostino, "Hertz's Researches on Electromagnetic Waves." *Historical Studies in the Physical Sciences.* 6 (1975): 261.

49. Charles Susskind, "Observations of Electromagnetic-Wave Radiation Before Hertz." *ISIS* 55 (1964): 32.

50. Hertz, *Electric Waves,* p. 2.

51. *Heinrich Hertz,* p. 215.

52. Ibid., p. 219.

53. Ibid., p. 215.

54. The story of how others missed making the discovery of radio waves is told by Charles Susskind, reference 49.

55. Hertz, *Dictionary of Scientific Biography.*

56. *Heinrich Hertz,* quoted on p. xxxv.

∩ 10 ∩
Curiouser and Curiouser

The price of gaining . . . an accurate theory has been the
erosion of our common sense.

Richard Feynman, *QED*

Thousands of years ago curious human beings asked why lodestone attracted pieces of iron. Until the nineteenth century the answers were very slow in coming. Then a rush of experimental and conceptual breakthroughs culminated in the development of the theory of electromagnetism, which showed that magnetism, electricity, and light were intimately related. The quest for an explanation of the fascinating force exhibited by the lodestone culminated in the discovery that current electricity created magnetic fields. It was a logical step to consider that circular currents might be responsible for magnetism in the molecular realm in order to explain the properties of lodestone and permanent magnets, and in the earth's molten core to explain geomagnetism.

Early in the twentieth century, the atomic theory of matter was developed. It pictured atoms to be made of a nucleus containing positively charged protons and chargeless neutrons locked together in tight embrace. Negatively charged electrons, which were discovered in 1895, orbit the nucleus in a number of discrete shells at specific distances from the center. The force that held the electrons in their orbits was electromagnetism, which explained why oppositely charged particles were able to attract one another. This force also held molecules together.

147

To account for magnetism itself, the model of the atom pointed to a source for the moving charges (electrons), but in the case of magnetic materials such as bar magnets the electrons did not actually flow along the bar of iron: they spin on their own axes. The spinning electron produces its own magnetic field, referred to as its magnetic moment. If a metal has a suitable atomic structure, magnetic moments of the electrons reinforce one another to produce the magnetic properties of lodestone, for example. In other metals the magnetic moments cancel out and the material shows no inherent magnetism.

In conductors, some of the electrons in the metal are free to move when driven by an external source of energy, such as the voltage available in a battery. In the case of a television antenna connected to a TV set, the current is driven by the radio waves striking the antenna. The radio wave, which is one form of an electromagnetic wave, provides the force that drives the electric current into the television receiver.

Recently, electromagnetism has come to be recognized as one of the four basic forces in the universe. The other three are gravity, the weak nuclear force, and the strong nuclear force. The weak nuclear force acts over very short distances and controls radioactive decay. The strong nuclear force is also a short-range force and it keeps protons and neutrons tightly packed in the nucleus of an atom. At short range, the strong force overwhelms electromagnetism, otherwise the protons and neutrons in the nucleus would blast apart. Gravity is part of our everyday experience. We live in the gravitational field of the earth. Try to escape it by jumping up. Doesn't get you very far, does it? The moon is also locked in the earth's gravitational field, while the earth is itself locked in the gravitational grip of the sun.

The electromagnetic force, unlike the weak and strong nuclear forces, is felt over vast astronomical distances, which is why we can see distant galaxies, by means of the light (an electromagnetic wave) they emit. Light is one type of carrier of the electromagnetic force, the carrier that allows the force to be sensed over a great distance. Gravity also reaches across the universe, which is why vast clusters of galaxies can be held together by their mutual attraction.[1]

But how can a force act over a distance with no observable mechanism for doing so? This brings us back to the question that Faraday, Maxwell, Hertz, and many others struggled to answer. The old idea was that a luminiferous ether existed that somehow passed the force from point to point, like a tumbling domino falls against its neighbor to cause the next one to fall, and so on. But when experiments to measure the motion of the earth through this hypothetical ether failed to detect its presence, physicists were left with the concept of a field to help the imagination cope with the mystery. Although the field notion allowed for elegant mathematical descriptions of phenomena, such a field was never directly observed. Even if Faraday could imagine he could see field lines in his displays of iron filing patterns, the effect was essentially an illusion. Yet field theories were beautifully elegant and served the purpose of explaining electrical and magnetic phenomena. But the fields were not real in the sense that anyone could isolate and study them as separate entities. Obviously the field concept could not be the final explanation.

In the 1950s a new paradigm (point of view) emerged, one that gave a more elegant account of how it was that various forces can act over a distance. The theory came to be known as quantum electrodynamics (QED), and its inventor, Richard Feynman, would earn a Nobel Prize for his efforts.[2] The essence of QED is that the force between particles such as electrons can be explained by picturing the exchange *photons,* massless particles that carry electromagnetic energy.

Electromagnetic fields produce distinct consequences when they strike matter, and the magnitude of their effect can be measured. For example, the presence of an electromagnetic field can be directly sensed by human beings. When we stand in the field of electromagnetic radiation produced by the sun we see the light and feel the heat. Heat is a form of electromagnetic radiation—infrared in this case—whose wavelength is longer than that of light but shorter than radio waves. We can literally feel the warmth of solar heat radiation striking our skins. We sense the energy in the wave after it has been converted into electrical signals in the nerve cells in our skin, which are then sent to the brain. Similarly, the force inherent in light interacts with matter in our retinas and is manifested as energy, which is converted into electrical signals that stimulate the brain to give us the experience of sight.

Force fields contain a certain amount of energy that can be directly sensed by our nervous systems, in the case of light and heat, or detected by experimental apparatus, in the case of other forms of electromagnetism, such as radio waves, ultraviolet, X-rays, and gamma rays. Although the energy contained in a light wave is trivial by comparison with physical actions such as closing a door, it is enough to cause a reaction when the light wave strikes the retina. The modern explanation for such electromagnetic radiation is more peculiar still; all forms of electromagnetic radiation can be treated either as a wave or as a particle. If light is a continuous stream of particles (as Isaac Newton imagined it to be), the problem of how it travels from point A to point B is easy to deal with. Obviously particles can travel through space. No carrier is required, no ether, because the particles just zip right along. The particle point of view avoids the problem of invoking a medium through which waves have to travel to traverse empty space. Yet both the wave and the particle concepts are useful.

In pursuit of a deeper understanding of the nature of the basic forces of the universe, we are now in a conceptual world where intuition is no longer a guide. Electromagnetic radiation, whether light or radio, X-rays, ultraviolet, infrared, or gamma rays, may travel as waves (as Hertz found for radio), or as particles (as Newton believed) called photons. This has been shown in many experiments whose details do not concern us except to note that you can never do an experiment that proves that light is both a particle and a wave at the same time. It is one or the other, depending on the experiment. This is odd because it suggests that we, the observers, determine the nature of light. Physicists have learned to live with this peculiar aspect of nature.[3]

The energy carried by the photon depends on its wavelength. That also means that its wavelength depends on its energy. The longer the wave, the less energy the photon carries; the shorter the wavelength the more energetic the particle. The shortest electromagnetic waves, gamma rays and X-rays, carry so much energy that when they hit a molecule in your body they can severely damage it, even to the point of causing dangerous mutations that may, in turn, lead to cancerous consequences.

The important point to bear in mind is that the *energy* of the electromagnetic *field* is carried by means of a particle or a wave, depending on how you look at it. After Hertz discovered radio waves, he showed that they traveled around his laboratory at the speed of light, both in the wires of his apparatus as well as through the air. In principle, he could have done an experiment to show that photons (a term used for all electromagnetic particles, even if they are not related to light) were involved. In practice, to do so would have been beyond his means.

Now let us return to our original question: What is magnetism? And why is there a distinction between electromagnetism and plain old magnetism? If the underlying force that links electricity and magnetism is electromagnetism, why do we observe what appear to be two distinctly different phenomena, electricity and magnetism, in everyday life?

The answer lies in the way we perceive or experience these two forces. In the theory of electromagnetism, electricity and magnetism are two aspects of an all-encompassing force, one phenomenon. Under normal conditions they are so intimately linked that we cannot speak of one without the other. In the parlance of physics they are said to be symmetrical. If we look at an electromagnetic wave from an appropriate point of view, such as is defined by a mathematical description, for example, the symmetry is perfect. To discover that an electromagnetic wave consists of two parts you have to destroy its cohesion; that is, break its symmetry.

Consider a perfect apple. If you want to learn about the apple you might take a knife and cut it in half. Now you have broken the apple's symmetry, and each half, while resembling the other, is no longer symmetrical in a way that defined the unblemished apple. Instead you have two halves, which may look alike, but the perfect apple no longer exists. Your curiosity about the apple destroyed its perfection. Even if you push the two halves together again you cannot restore the symmetry that existed when the apple still hung on the tree.

In recent years physicists studying the nature of the forces and particles in the universe have explored in great detail the concept of symmetry. It turns out that they can describe the nature of electromagnetism in terms of equations that exhibit symmetry. Under certain conditions that symmetry is broken and gives rise to electricity, on the one hand, and magnetism, on the other.

When does symmetry break?[4] The answer depends on who is asking! In nature the act of touching or of directly perceiving destroys the symmetry of the phenomenon you wish to study. To touch is to destroy. However, this does not prevent physicists from constructing an accurate picture of what the aforementioned apple must have looked like before it was cut. This is the case for electromagnetism. Electromagnetic waves (from the sun, and from radar, radio, and television transmitters) exist in space around us. When solar radiation strikes our hands it is felt as heat. When sunlight strikes our retinas it is sensed as light. When radio signals from a nearby transmitter enter the antennas of our television sets we can see a picture on the screen. In all cases, the symmetry of the original electromagnetic field is broken in the act of perception.

One side of the symmetrical coin, the electrical aspect of the wave in these examples, produces a change in our cells or in the amplifier of the television set. The electromagnetic wave from the transmitter triggers a wave in the antenna, and its electrical component is detected and amplified by the electronics of the receiver to produce the sound and picture carried by the signal transmitted from the television station. Similarly, it is the electrical nature of the electromagnetic wave called light that we experience through the mediation of our eyes as sight. The act of seeing the light breaks the symmetry of the electromagnetic field at your eye.

But so what? you may ask. What does all this have to do with the question asked by William Gilbert and Peter Peregrinus? What is magnetism? The new answer is that magnetism is the phenomenon manifested when the symmetry of the electromagnetic force is broken. The same symmetry breaking can also produce electricity. It depends which aspect of the broken symmetry you study in an experiment.

To have a static magnetic field, such as is produced by a lodestone, symmetry is already broken. If you dig down to the molecular level, you will find movement in the form of tiny currents giving rise to the magnetism. Those are not alternating currents such as are involved in producing electromagnetic waves. They are *direct* currents that perpetually flow in one direction in small loops. Direct currents are not symmetrical; they flow one way only. Symmetry would require that they spend as much time flowing one way as in the opposite direction. If they do so you have an alternating current and it immediately radiates electromagnetic waves.

If the electrons inside some substance were all rushing around randomly, the net magnetic field produced would be zero. A single particle moving in a circle would produce a field, but unless the particles move in an organized fashion their contributions to a net static field cancel out.

If we set up an electrical circuit and use a battery to send currents flowing one way only, we have defined a direction. A single direction is not symmetrical. A current that alternately flows one way and then in the opposite direction, and does so regularly, is symmetrical and it is able to generate an electromagnetic wave that travels through space.

The frequency at which the current changes direction determines the frequency of the radiation. An FM radio signal received at 100 megaHertz on the dial is produced by electrons rushing back and forth at 100 million times per second in the transmitter.[5] In the case of Hertz's experiment, he inadvertently produced radio waves at a frequency of about 30 megaHertz. The power in electrical outlets in United States homes arrives at an alternating frequency of 60 Hertz, which produces electromagnetic waves with a wavelength of 500 kilometers.[6]

The notion of symmetry finds its roots in elegant and beautiful equations used to describe electromagnetic waves. But why does the force of electromagnetism exist to start with? How did it come into being? Experiments and theory have shown that under conditions involving very high energy, found only in particle accelerators or in the early universe, two of the four forces, electromagnetism and the weak nuclear force, were one. Electromagnetism and the weak nuclear force were once a unity, two aspects of what is called the *electroweak* force.

The properties of the electroweak force were theoretically predicted and then observed in particle accelerator experiments in which matter is smashed together with such violence that energy conditions similar to those in the early universe are duplicated. Under such conditions the electroweak force was wrapped in symmetry. Such experiments allowed the properties of the force to be measured and they fit the theory.

When the universe was only 10^{-12} seconds old (one-trillionth of a second), it was very hot and very dense and filled with energy. Back then there was no electromagnetic radiation, no light, heat, or X-rays. Instead the electroweak force existed, and it had within it the potential to create two forces (the weak and electromagnetic force). But that could happen only after the universe cooled as it aged beyond 10^{-12} seconds. Then the symmetry of the electroweak force was broken, and the two sides of the electroweak force became manifested, one acting over a very short distance (the weak force), and the other over great distances (electromagnetism). The weak force resides deep within the nuclei of atoms and is important in controlling how bits and pieces of radioactive nuclei are held together and break apart. And because the electromagnetic force acts over vast distances, we can see stars at night.

Symmetry breaking is something like a phase transition that occurs when, for example, water is cooled enough to turn to ice. In the liquid state the water has certain properties and no preferred directions; it is symmetrical in a sense. But ice is crystalline. It has clear structure, which is not symmetrical in the sense that water is. The symmetry that existed in the liquid state is destroyed when water turns to ice.

Cooling in the early seconds after creation of the universe caused the two sides of the original cosmic coin, the electroweak force, to become manifested as electromagnetism and the weak force. These continued to exist independently and forever, as they do to this day, fifteen billion or so years later.

This discussion does not imply that all the electromagnetic radiation we now observe was created when the universe began. Instead, the laws of physics that describe electromagnetic phenomena were laid down in the first fraction of a second of the universe's existence. Similarly, the laws that determine the action of the weak force were created at that instant. Before then neither the weak force nor the electromagnetic force existed. In other words, when the universe was very young, hot, and dense, the laws of physics were different. More specifically, the nature of symmetry was such that electromagnetism and the weak force had not yet parted company.

As the universe cooled, symmetry breaking released these two genies from one bottle. The weak force took over control of the nuclei of atoms, and its strength determined the properties of matter. At the same time, the electromagnetic force became master of long-range connections between matter located everywhere in space. Electromagnetism communicates over great distances by means of massless particles called photons, or it involves waves, depending on which picture one prefers, and that may depend on a given experiment! Electromagnetism is also inherently weak, which is why we require enormous radio telescopes or large optical mirrors to gather faint whispers of radio or light energy from the depths of space.

But what is the mechanism that allows the electromagnetic force to be sensed over a great astronomical distances? And why is the weak force, its twin, only sensed over nuclear scales? Here a picture known as a Feynman diagram, shown in Figure 10-1, helps the imagination.

Consider a simple case of two electrons approaching one another. Like charges repel, so they will shy away from one another. But how do they do so? Feynman asked us to imagine that when the two electrons approach each other, one of them emits a particle, a photon, which then heads toward the other electron. When the photon is emitted, it gives the electron a kick backward, to cause the first electron to swerve away from the other. (This relates to a fundamental discovery by Newton, that an action always causes an equal and opposite reaction). When the photon strikes the other electron, it is absorbed and causes the second electron to change its direction of motion. The particle that carries the force is called a virtual photon because it cannot be seen by the observer. Through its influence it appears as if the two electrons are avoiding each other. The continual exchange of virtual photons determines the trajectory of the electrons relative to one another. The force invoked to describe this avoidance interaction is an aspect of electromagnetism. Feynman's model suggests that the force is carried by

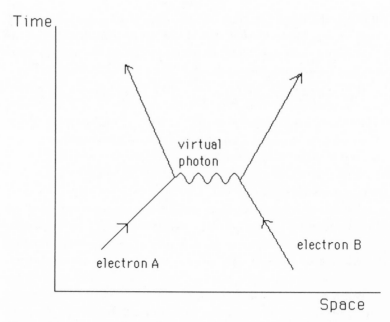

Figure 10–1. A Feynman diagram showing the electromagnetic inter-action between two electrons, which here repel one another at close ap-proach.

photons, which are massless. The force behaves *as if* a virtual photon is exchanged in the process of electrons repelling one another.

This model (QED) began to gain a foothold in 1950 and was found to be a very elegant way for describing the action of electromagnetic forces. Field theory, although still useful as a mathematical tool, then lost much of its power. At this point it is worth quoting Feynman, who cautions us that "the way we have to describe Nature is generally incomprehensible to us."[7] In other words, as we delve deeper into nature's secrets we do find answers, but those answers are not necessarily in a form that is readily comprehensible in terms of everyday language. That is why mathematics is not a common language like French, Japanese, or English. It is, among other things, the language of physics. Once the efforts of physicists to interpret what had been found in terms of common language failed, they were forced into the realm of what was not at all obvious or intuitive. But whoever expected that evolving concepts of nature's mechanisms would remain in the realm of human intuition?

A Feynman diagram can be used to show how we see a distant star. The surface of a star emits countless photons that traverse space and strike our eye. These massless particles could have traveled from the very edge of space-time before striking our retinas. Because they are massless they can travel forever without losing energy through close encounters with matter that would otherwise drain their energy. Of course, if a photon were to strike something on the way here, such as a planet or a dust particle in interstellar space, it would be absorbed and we would never see it. But since space is very empty, most of the photons get through and bring messages to earth about distant galaxies and the early universe.

In the case of the weak force, a similar phenomenon happens. Interactions at the nuclear level occur over a very short range and it was predicted that the particle that carries the weak force (the analogy to the photon) had to be very massive, which would explain why it didn't get very far before it ran out of steam. Such particles, called bosons, were discovered experimentally in 1983. That earned Carlo Rubbia, the leader of the team that did so, a Nobel Prize.

Now consider the other two forces of the basic quartet. The strong force also involves the exchange of particles. The carrier of the strong force is called a gluon, the hypothetical glue holding atomic nuclei together. It is heavier still, which means that the strong force can act only over very short distances. Similarly, it is hypothesized that particles called gravitons play a parallel role in holding matter together. In other words, what holds you against the earth is a continual exchange of gravitons between your body and the planet. However, gravitons have yet to be observed directly.

In summary, we can imagine fields as involving the exchange of particles that carry force. The bosons involved in the weak force are so heavy they cannot travel very far. That sets a limit on their action to the dimensions of the atomic nucleus. The boson's counterpart, the photon, carries the electromagnetic force. The photon is massless and is capable of traveling across the universe without interacting with matter.[8] That is why we can see galaxies located at distances as vast as ten billion

light years. These two forces are the two sides of what was originally the electroweak force. When the symmetry was broken,
bosons and photons became part of the vast realm of possibilities that define our physical universe.

Now we return to the Kantian view that the forces of nature
are part of some overall unity. The unification of the forces,
the linking of gravity and electromagnetism, was a subject dear
to Albert Einstein's heart. The challenge to unify all four forces
under a single umbrella has attracted many famous physicists.
Cranks and pseudoscientists are also drawn to the subject like
moths to a flame. For some reason, the temptation of finding a
unified field theory has the power to lure rational and irrational
minds with equal force.

The original goal in the search for a unified field model was
to link gravity and electromagnetism in one fell swoop. It was
considered to be the Holy Grail of physics and many knights in
shining armor ride large research grants in the quest for the
answer. The path to success is now visible.

The strong nuclear force is not part of our direct experience. This force acts as the powerful glue in atomic nuclei
and is extremely strong at very short distances, but it is not felt
beyond the nuclei. We have also seen that electricity and magnetism are subsumed under electromagnetism in one force and
have noted that under the high energy conditions found in the
early universe, or in particle accelerators, electromagnetism and
the weak force behave as one—the electroweak force. In following the path toward a fuller understanding of how the other
forces are linked to these (the strong nuclear force and gravity),
we rejoin the trail at the point where the electroweak force is
in a state of symmetry, when the universe was less than 10^{-12}
seconds old. Extrapolating even further back in time, back to a
condition when the universe was more compressed and far hot-
ter, physicists discovered that their equations showed that a new
type of symmetry played a role. The electroweak force is apparently the "twin" of the strong force. Under the extreme conditions that existed when the universe was less than about 10^{-38}
seconds old they were one. This short time span is incomprehensible to our minds, yet it was important in the evolution of
the universe. In that brief interval after the universe came into
existence in the Big Bang, the electroweak and the strong forces

were part of a symmetrical whole, a higher-order symmetry. Physicists have described this state under the umbrella of GUTs: Grand Unified Theories.

At an age of 10^{-38} seconds the universe was only 10^{-33} centimeters in size. Before that moment there were only two forces present, gravity and another that contained within it the seeds for the strong, weak, and electromagnetic forces. To explain this, physicists invoke a state of greater symmetry, known as supersymmetry, that existed in a *multidimensional* universe that must have existed when it all began. In order to account for supersymmetry many more dimensions than the three of space and one of time to which we are accustomed in our lives are required. The ultimate parent apple representing nature's basic forces may have existed in a universe with perhaps twenty-six or ten dimensions (according to two out of a variety of current theories that have yet to withstand the test of time themselves).

But why stop there? Why not look even further back in time to explore whether gravity, too, was once unified into one all-encompassing scheme? A search for this level of unity involves the Theory Of Everything (TOE). At an earlier instant in time, before 10^{-38} seconds, when the universe was even more unimaginably tiny, hot, and dense, all the forces were probably one. Unfortunately (or is it fortunate?) it is impossible to produce laboratory simulations of what the universe was like then, although it is possible to develop theories to predict what should be observed in later epochs and up to the present time.

A Theory Of Everything would allow all the forces to be accounted for in a single model and will provide a broad description of the origin and evolution of matter, galaxies, stars, and planets, all of which emerged after the hierarchy of symmetry breaking was complete. This may well lead to a theory of *everything* whose compass may include our existence on earth, given that we may learn to explain the complexities that result from combining simple matter into higher-order structures (molecules, cells, organisms) in sufficient detail. Perhaps it is only a matter of time for creative intellects, which have emerged out of the sea of matter and forces that constitute our universe, to find the explanation.

The point of touching so briefly upon these topics at the frontiers of modern physics is to indicate a remarkable impli-

cation of the existence of the new theories that have been invoked to explain the existence of matter, as well as the forces
between various forms of matter. For the first time it appears
that humankind, as represented by its physicists at least, is on
the brink of understanding everything.[9] This does not mean
that the details will be understood; only that the broad picture
that underlies our existence will have been revealed.

The issue is not whether a complete, self-consistent picture,
the Theory Of Everything, will emerge in the 1990s or in the
early part of the twenty-first century. Instead, it is that a possible resolution of many fundamental questions concerning the
physical universe may someday, relatively soon, be answered by
a comprehensive theory. At no previous time in history could
this have been claimed with quite the enthusiasm that is now
devoted to the quest for the unifying concepts. Ampère, Faraday, Einstein, and many other physicists have always sought to
unify their understanding of the forces of nature. None of them
succeeded. With the marriage of electromagnetism and the weak
force, the first major step to a unification of the four basic forces
was taken. Final success is only a matter of time. The path toward
that unification is visible, and all physicists have to do is move
ahead systematically and carefully.

Today we live in a universe in which four forces seem to
control matter. Now we know that these forces emerged as separate entities when the universe was a milli-, micro-, micro-,
micro-, micro-, micro-, microsecond old thanks to a hierarchy
of symmetry breaking. (For those readers familiar with some of
the concepts of modern cosmology, the time when the forces
were unified was before the era of inflation. Inflation began
when symmetry was first broken.)

The British physicist Stephen Hawking has made an optimistic statement about whether the discovery of the Theory Of
Everything will be possible. "I believe there are grounds for
cautious optimism that we may now be near the end of the
search for the ultimate laws of nature."[10] Skeptics react to this
statement and suggest that people have always thought this. But
we are hard-pressed to find historical examples where such a
consensus existed about the likelihood of success.

As far as our story is concerned, the search for the answer
to the question of what is lodestone has led us via action at a

distance through electromagnetic fields to QED, GUTs, and a possible TOE. The answers to many related questions now find a common answer in the remarkable properties of the physical universe in which we find ourselves.

It is awesome to confront the fact that the human brain, by asking questions, has found answers. These answers work so well that we are left with the stunning awareness that we live in a universe in which at least one form of consciousness has emerged that has the capacity to comprehend not only its own existence but its origins as well.

NOTES

1. By observing how many galaxies of known mass gather together in certain clusters it is possible to calculate the strength of their mutual pull. The galaxies themselves are moving with respect to each other, and the problem referred to in the text is that their energy of motion appears to be too large to allow gravity to hold them close to one another for very long. Thus astronomers have argued that there must be "missing mass," that is, some form of matter, as yet undetected, holding the clusters together. The search for this mystery mass is one of the great challenges in astronomy and particle physics today.

2. For a popular level exposition see Richard Feynman, *QED: The Strange Theory of Light and Matter.* (Princeton: Princeton University Press, 1985), p. 119.

3. Most books dealing with modern physics at the popular level discuss this in more detail.

4. The role of symmetry breaking by an observer is beautifully discussed by F. David Peat, *Superstrings and the Search for the Theory of Everything.* (Chicago: Contemporary Books, 1988).

5. A Hertz is the name given to the unit of frequency of one cycle per second.

6. From the relationship that for an electromagnetic wave frequency times wavelength equals the speed of light; $f \times \lambda = c$.

7. Feynman, *QED*, p. 77.

8. This statement is not strictly true, as Einstein showed. A strong gravitational field bends light trying to sneak past, but a discussion of this is beyond our scope. Black holes result when the gravitational pull of a massive, small object is so strong that the photons are pulled into

the object and it disappears from view and makes the structure invisible; hence the name, black hole. Under such extreme conditions the laws of physics break down.

9. Stephen Hawking, *A Brief History of Time*. (New York: Bantam Books, 1988) for an optimistic appraisal of how close we are to a full understanding.

10. Hawking, *A Brief History of Time*, p. 156.

n ll n
What If?

What must nature, including man, be like in order that
science be possible at all?

Thomas Kuhn,
The Structure of Scientific Revolutions

PERCEPTIONS about the
nature of magnetism have
changed dramatically over time in a systematic manner. Superstitious beliefs held sway for thousands of years and slowly began to make way for enlightened thinking based on experimental studies of the phenomenon. Experiment, however, could
begin in earnest only after sensitive and sophisticated measuring instruments were built.

Once magnetism began to be understood in a physical sense,
the concept of action at a distance became popular, and the
force of magnetism was recognized to be directly related to the
properties of the magnets themselves. Then Faraday's insights
led to explanations for magnetism in terms of the existence of
invisible fields. Physicists began to deal with the fields as entities
in themselves. The characterization of a field was independent
of the physical properties of the magnets.[1] A critical aspect of
this view was that field theories included time as a factor. The
magnetic force, for example, was not propagated instantaneously through space but traveled at a finite speed—that of
light. Previously, action-at-a-distance descriptions, whether the
concept applied to the force between masses, charges, or magnets, had invoked instantaneous propagation and overlooked
the time element. Such descriptions had been statements about

the magnitude of the forces.[2] Field theory had its limitations, because a field could never be observed directly, even if the notion gave rise to the wonderfully effective model that Maxwell used so successfully to describe electromagnetic phenomena.

A paradigm shift occurred with the invention of quantum electrodynamics (QED). It avoided fields and instead pictured the exchange of virtual photons in allowing force to be sensed over a distance. QED has been described as "unquestionably the most successful theory in the history of science."[3] Yet QED was in turn superseded by the so-called standard model, which in the 1970s and 80s unified electromagnetism with the weak force of nuclear physics. The result of marriage was the electroweak force. The standard model considered symmetry as a basic issue and that provided an understanding of fundamental particles and their interactions. Then, in the late 1980s the search for the Theory of Everything began. It would, it was hoped, link the electroweak force and the strong nuclear force (already related in Grand Unified Theories, or GUT) with gravity into a single underlying order of nature, the Holy Grail of physics.

The perspective just outlined implies that a great deal of progress in understanding the most basic aspects of physics rested on the intellectual impetus offered through the natural occurrence of lodestones in the environment. The presence of nature's magnets triggered a long-lasting and persistent train of questions about the nature of physical phenomena, without which, in the time that elapsed, we would not know as much about the working of our universe as we do. In this context I want to play with some new ideas, in particular the exploration of consequences if either lodestone had not existed or if the earth were not magnetic. More specifically, how far would science have progressed if natural magnets had not been available to pique human curiosity?

A question such as, What would have happened if the earth were not magnetic? is of a type that scientists confront with a considerable amount of schizophrenia. On the one hand, in order to design an experiment, "What if" questions—for example, What will happen to the gas in the flask if the temperature is raised?—are perfectly legitimate and necessary. On the other hand, asking a "What if" that cannot be tested by experiment is regarded by many of my colleagues as a cardinal sin. Speculations that concern situations where no experiment can be designed to test for an answer are simply regarded as bad science. Anyone who dares ask such a question is regarded either as a crank or in immediate danger of losing his or her senses. Examples include, What will happen if we go faster than light? and What if there is a parallel universe in which people live who are mirror images of us? These are in a category that reveals that the questioner doesn't appreciate the nature of science. To answer such questions is not within the realm of physical possibility, so why bother to ask them?

The point, of course, is that scientists can observe the universe only as it is, not as it might have been under different circumstances. Despite this caution, I will approach two forbidden questions, because to me they are fun to play with, especially when we perform thought experiments with the new concepts.

My first question is this: What would have happened to our technological development if the earth had not been magnetic? Since we have not yet discussed why the earth *is* magnetic, let us deal with that first. Research into the nature of magnetism became possible only after Gilbert cleared the decks of superstition about lodestone and recognized that there were interesting questions to be answered through experimentation. Superstitions had served to provide answers for two millennia, but what did those superstitions concern? They attempted to answer the question, What gives lodestone its peculiar property? The modern answer is that magnetite, which is what lodestones are made of, is naturally magnetic because its particles respond to the pull of the earth's magnetic field. When magnetite was initially deposited, its particles were aligned by the gentle tug of the terrestrial field and the lodestone retained a memory of that ancient field's presence. This same phenomenon of magnetic memory allows us to store sound and video on cassette tapes. Similarly, when lava flows out of the earth and cools and solidifies, its molecules experience the pull of the earth's field, which orients them in the solidifying rocks to reflect a net residual magnetization. This can be detected by sensitive magnetometers. The temperature below which the iron or the lava has to cool to lock in the magnetic memory is called the Curie point.

Conversely, if one heats magnetic material such as lodestone above the Curie point it will lose its magnetism, something first noticed by Gilbert, although the name of the French chemist Pierre Curie (1859–1906) was adopted in honor of his work in studying the phenomenon.

The upwelling of material from the earth's core along the midoceanic ridges continues in a systematic manner, and the sea floor contains structures running parallel to the ridge in which the magnetic field direction reverses with time. Reversals occur on average about once every 200,000 years. I will return to the reversals later.

Lodestone is magnetic because it was laid down in a magnetic environment; its molecular structure is such that the alignment of particles was great enough to make lodestone noticeably magnetic to miners over two thousand years ago. This then raises the interesting question of what would have happened in human history if the earth's field had not existed, everything else being equal. One thing we can state unequivocally is that no matter how lodestone (magnetite) particles came to be deposited, their molecules would not have lined up in an organized pattern and humankind would not have had a supply of natural magnets with which to play.

But why is the earth magnetic? Widely held opinion is that deep within the earth's core hot, molten magma rises, cools, and sinks, only to be heated and rise again. The heat comes from natural radioactivity, which has kept the planet's interior hot since the earliest days of the solar system. With passing time the radioactive sources expend their energy, and so this fountain of heat is slowly wasting away.

Within the rising and falling masses of magma, which can conduct electricity, the rotation of the earth creates organized patterns of circular motion called eddies. As a result, the interior of the planet acts as a giant dynamo within which circular currents produce a magnetic field. At the surface of the earth the strength of the field is about 0.5 gauss (a gauss is a unit of magnetic field strength named after Carl Friedrich Gauss [1777–1855]). Although the terrestrial dynamo theory is broadly understood, the details remain a mystery. For example, geophysicists have a terrible time trying to explain why the field reverses from time to time. That happens on average once every

200,000 years or so, although it hasn't happened during the last 800,000 years. Somehow the dynamo switches off and then on again to create a field in the opposite direction. During the transition from one direction to the other, there may be a period of time, perhaps lasting thousands of years, when the field is essentially zero. This state would have very serious consequences for life on the earth's surface, since the field acts as a protective umbrella that shields us from the particle stream that continually blows out of the sun: the solar wind. Under normal conditions the solar wind is deflected around the earth by the magnetic field, which reaches high above the surface into surrounding space, much as the shield of the fictional starship *Enterprise* in the television science fiction series "Star Trek" deflects photon torpedoes that threaten to destroy it. If solar wind particles (mostly protons and electrons) were to strike the earth's atmosphere directly, they would trigger a series of complex chemical processes that would destroy ozone.

Ozone in the high atmosphere (stratosphere) absorbs solar ultraviolet radiation, which is harmful to life. This layer is currently the focus of much public attention because we are damaging it so badly. But without the protective magnetic field, the solar wind would impact the atmosphere directly, alter its chemistry, and remove the ozone layer virtually completely. Many species live on the knife edge of tolerance to ultraviolet radiation. A field reversal involving a period of zero field, as the field drops from its maximum value pointing north–south, say, before reappearing to point south–north in the geographical sense, would be devastating for life on earth. The geological record shows that at times of field reversals many different species did become extinct. But the details of this scenario are far from understood.

Geophysicists do not yet understand why or how the field reverses. Does the field drop to zero and then reappear after some time, perhaps after thousands of years, or does it gradually swing from one direction to the other without dropping to zero in between? These two alternatives would have very different implications for life, should a reversal happen in our time. In the first case it would likely be fatal for us, whereas in the second case it may merely be very inconvenient for navigation. Progress in understanding what actually happens is being made in a dramatic manner by Robert Coe of the University of California at Santa Cruz and two French colleagues. He studied the magnetic field record in lava that flowed from an Oregon volcano during an actual field reversal event fifteen million years old. The field direction apparently fluctuated wildly from day to day. This raises the specter of a field reversal being possible at almost any time, as opposed to occurring slowly so as to give us warning that it might happen. For example, the long-range measurements of the terrestrial field strength, which have only been possible at the desired level of accuracy for the last thirty to fifty years, indicate a decreasing field that would drop to zero in a thousand years or so. But this decrease may only be a temporary decline. Only time will tell. The only thing we can be reasonably sure about is that if the field were to drop to zero and remain there for months or years, the impact of the solar wind on the ozone layer would dwarf the destruction we fear from our chemical pollution of the stratosphere.

This perspective allows us to answer the question of what would have happened if the earth had never had a field, as is the case on Venus, otherwise regarded as earth's twin planet in terms of size. Without a field to act as a shield, the solar wind would impinge the atmosphere and no ozone layer could have been sustained. That means that life of the complexity and variety we find here would never have emerged. Instead, primitive organisms, such as existed in the early stages of earth's existence four billion years ago, would probably still be dominant. Four billion years ago there was no oxygen in the atmosphere and hence no ozone either, yet the planet was teeming with primitive organisms, which could not begin to evolve in the direction of the type of life we know today without the creation of an ozone layer protected by a magnetic field.

The terrestrial magnetic field is therefore essential for the origin and sustenance of life as we know it. Without the field we would not be here. But now imagine our planet with a field but with no lodestone to make known the existence of magnetism in the first place. How would scientists have learned about the presence of this force without lodestone's help? For that matter, why does lodestone exist, and why is only some iron in iron mines laid down in the form of magnetite?

Let us consider the first question first. Without lodestone,

ancient mariners would have had no compasses and early experimenters such as Peregrinus and Gilbert would have had
nothing with which to experiment. Since Gilbert's book on the
magnet was the world's first scientific treatise, that honor would
have had to go elsewhere; who knows where and who knows
when. Surely it would have happened well after 1600. People
might still have been fascinated by static electricity produced by
rubbing amber, but the magical properties of amber were less
dramatic than those of lodestone. You had to do something to
amber to make it work, whereas lodestone worked without anyone having to do anything to it. Magnetism was more mysterious because it existed whether or not anyone touched the lodestone.

In the absence of lodestone, at what point would anyone
have learned about the existence of magnetism? This question
is, of course, impossible to answer, but I think we can safely
speculate that it would have been very long *after* anyone actually did learn about it. Without lodestone, famous physicists
such as Oersted, Ampère, and Faraday would have had no
magnets in their laboratories and could not have learned about
electromagnetism when they did. In that case, twentieth century civilization would not have reached its present degree of
technological evolution. I would go so far as to argue that without lodestone to trigger the curiosity of long-dead philosophers
and scientists our culture might still be getting around on
horseback.

History teaches us that research into the nature of electricity
and magnetism proceeded hand-in-hand and led to the invention of the electric dynamo and motor, both of which use magnets. Without natural magnets to stimulate research, such devices would surely not have been invented when they were during
the nineteenth century. The breakthroughs would have come
later, very much later.

Given the importance of lodestone as a cornerstone for our
technological progress, let us ask why it exists. Lodestone is found
in rock strata associated with iron deposits. Why do such strata
exist? Why isn't all the iron in such mines magnetic? Why is
any of it magnetic? The newly discovered answer to these questions, which have been debated for years, is quite extraordinary.

Lodestone is made by bacteria!

According to Derek Lovley and his colleagues at the US Geological Survey in Reston, Virginia, a sediment organism, a bacterium known as GS-15, survives in the absence of oxygen (it is anaerobic) and requires iron for its metabolism.[4] GS-15 eats iron for breakfast, lunch, and dinner. These bacteria convert common ferric oxide (the common form of iron in iron ore) to magnetite (lodestone particles), and over great periods of time in the pre-Cambrian world, a billion or so years ago, formed magnetite layers in iron formations. The Lovley team recently found such bacteria living at depths of 200 feet beneath the earth's surface. The lineage of these microorganisms may date back to an era before photosynthesis even started to develop. Magnetite deposits have also been found in the oceans, where they range in age from the Quaternary (two million years ago) to the Eocene (fifty-five million years ago). These layers have also been laid down by bacteria.[5]

Now we can answer another "What if" question. If it were not for the bacterium GS-15 we would not have radio and television today.

The discovery that GS-15 plays an important role in the creation of magnetite deposits removed from the center of the picture another type of bacteria, which uses magnetite to help it find its food supply. The phenomenon is known as magnetotaxis, referring to the ability of a creature to respond to the earth's field. Magnetotactic bacteria have little compasses in their bodies, compasses that aren't used for telling north from south, but up from down.

The existence of magnetotactic bacteria is a wonderful example of nature's magnificent obsession. Life expands into every possible ecological nook and cranny and makes use of all the opportunities offered to survive. One such niche involves using information contained in the terrestrial magnetic field. As Robert Norman discovered back in 1581, the earth's field is inclined to the ground. Today we know that this is because the earth is a giant magnet and the field lines point toward the poles inside the planet. Therefore, everywhere on the earth's surface the field is inclined to the horizontal, except at the magnetic equator. Near the magnetic poles the field lines point straight up and down, and there magnetic compasses are of no use.

In 1975 Richard Blakemore[6] at the Woods Hole Oceano-graphic Institution found magnetotactic bacteria that swam along the earth's magnetic field lines.[7] In laboratory experiments, when the field was reversed the magnetotactic bacteria swam in the opposite direction along the field lines of the laboratory mag-net. These bacteria normally live in lakes, swamps, and marshes and thrive on nutrients found in sediments at the bottom. In order to find their food, they swim down along the earth's mag-netic field until they reach their grazing grounds. These organ-isms evolved to do this thanks to magnetite grains in their cells that are linked in long chains to form a rudimentary field de-tector. A slight pull from the North Pole of the earth, for ex-ample, guides them to swim in that direction, which takes them down into the sediments. If you take a Northern Hemisphere magnetotactic bacterium south of the equator, it would con-tinue to swim along the field lines. But in the Southern Hemi-sphere the field is headed up and out of the ground. North appears to be up! The bacterium following the field would find itself at the surface of the water and die from starvation. Lab-oratory experiments have confirmed this.

What happens to such bacteria when the earth's field re-verses? Nature has inadvertently taken care of this as well, thanks to mutations. In any normal population of bacteria there will be aberrant mutants who swim the wrong way and expire. But when the field reverses it is they who have the survival advan-tage. Then the normal ones find themselves starving and the mutants survive to begin a new population and perhaps even a new species of magnetotactic bacteria. As I said before, what happens when the field is zero during a reversal is anyone's guess. If the new data on the wild fluctuations in field direction during a reversal are relevant, then we can imagine bacteria as becoming hopelessly lost as they swim this way and that, at least until the field settles down to its new configuration. For a while perhaps those that swim sideways survive!

The discovery of these bacteria led to speculation that vast colonies accounted for the occurrence of magnetite strata where lodestone was found. Two billion years ago magnetotactic bac-teria might have been the most prevalent species on earth.[8] To-day they are still widespread and have even been found in cer-tain soils (e.g., in Bavaria). They convert iron into magnetite in their cells[9] and that contributes to the widespread occurrence of magnetite grains in the soil. But these magnetite grains are very small and do not occur in sufficient numbers to produce magnetic rocks, even over geological times. Instead, GS-15 fits the bill in providing lodestone strata.

While on the subject of animals using magnetism to survive, many other species besides bacteria are known to be magnetically sensitive. These include honeybees,[10] pigeons,[11] bobolinks,[12] various migratory bird species, algae, tuna, and dolphins and whales.[13] The subject of biomagnetism is growing rapidly as scientists confront the fact that many animals have magnetite in the cell structure in specific locations of their brains, as is true for Pacific dolphins, for example. It is very likely that the dolphins also use the field to tell up from down. A dolphin is an air-breathing mammal and must regularly come up for air. In a deep night-time dive, how does it tell up from down? Its sonar may not reach the bottom or the surface of the water. Since the dolphin is neutrally buoyant it cannot expect to find up by relaxing and seeing which way it drifts. It seems very likely (to me at least) that being able to sense the pull of the earth's magnetic field in order to tell up from down would be an effective way to remind dolphins which way to swim to get more air.

Magnetic sensitivity in whales may help account for many stranding incidents. Certain whales use magnetism to navigate, and they have been known to lose their way when they come close to shore where local irregularities in the field point inland rather than along the coast. Sensitive mapping of magnetism over the whole world shows that the earth's field is not uniform but very patchy, and along the east coast of the United States, for example, there are many irregularities in the field. Correlations between the location of whale strandings and local magnetic field anomalies are high.[14]

Even humans are alleged to have a certain degree of magnetic sensitivity, if one believes a report that we have trace amounts of iron in our noses, which may be used as a rudimentary compass.[15] I am skeptical of this claim, because I have sat in a device capable of detecting magnetic fields of extraordinarily low levels of 10^{-12} gauss (one-trillionth of a gauss) and was shown that magnetic fields at this level were only produced by small currents just above and behind the ear. The device (called a SQUID) does not detect fields elsewhere near the surface of the body. Magnetite in our noses would have been easily detectable by this machine. Also, what evolutionary pressure would have provided *Homo sapiens* with magnetic sensitivity? We surely don't need it to tell up from down, and we use other natural phenomena to navigate (landmarks, sun, moon, stars, etc.).

Now let us return to the question, What would have happened without lodestone? recognizing that it can be reduced to, What if a certain type of bacterium that thrives on chewing iron had not lived on earth? More specifically, without lodestone, how and when would humans have discovered magnetism?

In the absence of lodestone, an era of superstition about its magic would never have occurred although the magical properties of amber would still have aroused curiosity. Without lodestone, William Gilbert would have pursued another hobby instead of writing a book called *De Magnete*. Columbus would not have found America when he did; his course across the ocean, everything else being equal, would have been different if he had sailed by the stars instead of a magnetic compass. Without lodestone, the great explorations of the world would have followed a completely different course.

Without lodestone, von Guericke might still have invented his electrostatic generator when he did, and Coulomb his torsional balance. Volta might still have invented the pile for producing electric current. But what then? Experiments on electricity would have continued, and we can assume that someday someone would have observed an odd effect when currents flowed in wires. But when, and which effects? Even if someone had noticed, the usual reaction was to ignore the new effect. For example, in 1820 Davy saw that iron filings clung to a current-carrying wire to form a mass ten or twelve times the thickness of the wire, but he paid little attention to the phenomenon.[16] Of course, he had iron filings lying around only because he was experimenting with magnets. We can be sure that good researchers would not have been dirtying their labs with iron filings to start with. If a few flecks of metal had stuck to a wire they would have been brushed off. In that same year François Arago showed that iron filings were magnetized by a current-carrying wire.[17] But again, he was experimenting with iron filings because he already knew about magnetism. Nearly simultaneously, Oersted announced that he had found that an electric current influenced a compass needle. He would not have found that without a compass needle to play with, and compass needles were made using lodestone.

As early as 1801 it was reported that Nicholas Gautherot, the French chemist, noticed that two wires carrying current stuck together. Although he reported this finding, he took little notice.[18] The same thing was observed by others experimenting with electricity and magnetism, and they also ignored it. All these scientists were aware of magnetism and were in pursuit of solutions to specific problems. It is easy for us to suggest that they would have taken greater heed under different circumstances and thus discovered magnetism, but discovery of a new phenomenon is extremely difficult. To repeat the words of Louis Pasteur, "In the fields of observation, nature favors only the prepared mind." What could have prepared the mind to recognize the existence of a totally unexpected phenomenon, which is what magnetism would have been to those unaware of natural magnets? As it was, those physicists who first saw that current-carrying wires attracted one another took little heed of the effect, and surely their minds should have responded.

No matter how often Oersted sent currents along wires, he never would have noted the magnetic effects because he did not have compasses around ready to be disturbed by the current-induced fields. Without magnets to play with, the beautiful symmetry between electricity and magnetism would have remained undiscovered for decades and perhaps even centuries, if the history of discovery is a guide as to how difficult it can be to perceive a new phenomenon. Perhaps if someone had wound long lengths of wire into coils and sent currents through them they would have discovered that pieces of iron were drawn into the coil, or that the coils attracted one another. But why would anyone have wound long lengths of wire into coils? Just for the fun of it? We cannot expect that physicists would have doodled with wires to see what happened when current was sent through a circuit in case it produced an unexpected force. Faraday and others wound such coils because they knew that currents produced circular magnetic fields and wondered whether circular currents would therefore create straight magnetic fields. This was an enormous step even for Faraday, one of the few willing to think about circles.

In Gilbert's time it was known that when molten iron was wrought into shape by beating, the metal became weakly magnetic, but I suspect this was recognized only because of interest in the power of attraction of lodestone. Perhaps physicists would have noted that pieces of iron were magnetized by lightning strokes. But how would they have known that? It was the change in the magnetic properties of compass needles that revealed the phenomenon. Random pieces of metal lying about might have been weakly affected, but no one would have noticed. Without lodestone in their arsenal, they would have had no compass needles with which to detect the magnetism in the iron.

Although we cannot guess when anyone would have discovered magnetism, without lodestones to set scientific minds on the right track our society would almost certainly have entered the twentieth century (all other things being equal) with no knowledge of magnetism. That implies no motors, generators, relays, no knowledge of radio waves or X-rays, no radio, television, radar, or computers with magnetic memories. (Without radar, World War II would likely have had a very different outcome.)

We can imagine that some eagle-eyed observer such as a Hertz would inevitably, albeit much later than in Hertz's time, have considered the attraction between current-carrying wires as worthy of study. How long would it then have taken to find an explanation for the effect in terms that did not rely on electricity alone? The mental step required to confront a new force of nature would be far from trivial, as we have already seen in our story.

There is nothing obvious that could have given us a clue about the existence of magnetism in the earth, and hence of magnetism in general, except the occurrence of auroras, great sheets of light that illuminate the northern or southern skies following solar flares, explosions on the surface of the sun. They send electrically charged particles past the earth's magnetic field where energetic interaction with the earth's magnetic tail can send some of the particles to stream back along the field lines to come crashing into the high atmosphere causing the magnificent glow of the auroras. This is understood because geophysicists are aware of magnetic fields and the existence of electrons, which, in turn, were discovered thanks to the combined knowledge of electricity and magnetism. Modern explanations for auroras rest on knowing the earth is magnetic. Without knowledge of magnetism, how would someone figure out that the auroras were telling us that the earth had a previously unknown property?[19]

The first inkling of the existence of magnetism might have come from astronomers. Sooner or later someone would notice that the sunlight originating near sunspots shows oddly different characteristics compared with light from the quiet surface of the sun. Spectral lines are broadened near the spots.[20] Arguments would have raged for decades about the phenomenon, because the splitting manifests itself as a small wavelength shift. At first astronomers might have inferred complex velocity patterns around sunspots to produce these shifts through the Doppler effect.[21] Then, sooner or later, someone might discover that the light from these different components was polarized.[22] But polarizing filters are only used on telescopes to observe effects on light due to magnetism. It is not obvious that anyone would have bothered to search for the tiny amounts of polarization had there been no reason to expect it.

The moral of this fiction is that if there exists an alien planet otherwise identical to earth but which, by some quirk of evolutionary fate, does not have a certain type of bacterium, their scientists might still be struggling to understand electricity, oblivious of the fact that it was one side of the coin called electromagnetism. That civilization might be technically advanced to some degree, but it would not know about radio waves and would have no radio and television. It certainly would not be able to communicate over interstellar distances.

The point of all this speculation is not to suggest that without lodestone the manifestations of magnetism would never have been recognized; only that awareness would have dawned much later than it did, and we would not have advanced—technologically speaking—as far as we have.

We began this chapter by suggesting that most "What if" questions are fruitless unless we can come up with ways to find answers. Clearly we have learned a great deal about electricity, magnetism, and electromagnetism, thanks to nature's thoughtful gift of lodestone. If you accept my arguments, you might wonder if, by analogy, it is possible that natural processes exist of which we are oblivious. But this is a "What if" we cannot deal with, at least not as respectable scientists. We have no grounds for framing such a question, except by analogy, and analogies have a way of being off the mark. No matter how fascinating the idea may seem, it is best left to science fiction for elucidation. We can only deal with what we know exists, with phenomena and processes that we can detect and measure.

Now consider why the human intellect was ever directed toward the study of electricity. The word electricity has its root in the Greek word for amber, *electron*. Amber is fossil resin that was used in making beads. It also originated in something that was once alive; in this case, trees. When rubbed, amber becomes electrically charged to the point where it can attract light objects in a manner very similar to the magical effect of lodestone.

> Great has ever been the fame of the lodestone and of amber in the writings of the learned; many philosophers cite the lodestone and also amber whenever, in explaining mysteries, their minds become obfuscated and reason can go no farther.[23]

So wrote Gilbert in 1600. If you couldn't explain something weird, blame it on amber, and lodestone! He noted that "over-inquisitive theologians" and medical men had been wont to do so for centuries. But, as Gilbert realized, that was to err.

> But all . . . are ignorant that the causes of the lodestone's movements are very different from those which give to amber its properties; hence they easily fall into errors, and by their own imaginings are led farther and farther astray.[24]

The Moors used amber in sacrifices and in the worship of gods. According to Gilbert it was "certain that amber comes for the most part from the sea; it is gathered on the coast after heavy storms."[25] It also entombs insects, which makes amber an object of attention even today. It was not just amber that attracted chaff. Many other substances, such as amethyst, diamond, and sapphire did the same. And all of them did so better "when in mid-winter the atmosphere is very cold, clear, and thin; when the electrical effluvia of the earth offer less impediment, and the electrical bodies are harder."[26] Here we enter the realm of everyday experience, especially in centrally heated buildings in the winter, when the scuffing of our shoes on large expanses of carpet may produce so much static that when we lean over to touch a water fountain a mighty discharge sends a jolt through our bodies. We only experience static charges if the air is dry. When it is humid we do not have this problem. Therefore, what would happen on a planet that is permanently humid? How would a civilization there discover electricity? Lightning would not help. Lightning was only recognized to be an electrical discharge when Galvani performed his notorious frogs' legs experiments and made the connection.

Perhaps intelligent life on a humid planet with no lodestone would find it nearly impossible to discover the existence of either electricity or magnetism, and would forever remain in awe of lightning. This speculation may seem particularly fruitless, but I think it contains a lesson. Not all planets in the Milky Way will be precisely like earth. There may be countless inhabited planets where local conditions cause those civilizations to think completely differently from *Homo sapiens*. Surely that is more likely than expecting large numbers of twin societies, which the modern myth of "Star Trek" has projected into our unconscious. Awareness that a planet slightly different from earth would support a very different array of species has a profound bearing on our speculation about the possible existence of extraterrestrial intelligence. As Stephen Jay Gould has argued, if nature were to replay the tape of life, the probability that *Home sapiens* would again appear is incredibly small.[27] Unless conditions are very close to terrestrial, we might expect technological civilizations on earthlike planets to be rare indeed.

Our explorations of nature can only be guided by what strikes our fancy, and what strikes our fancy is what strikes our senses. Thus gravity was always a candidate for research because it affects every one of us. It was difficult not to notice that if you tripped you fell. Why fall? Sooner or later someone would ask, "Why do I always fall down when I trip? Why don't I fall up?"

Similarly, there are phenomena in our environment that occur as the result of human action, which lead curious and inquiring minds to ask questions. I recall that in my school days I read what must have been an apocryphal story about James

Watt, who stood watching a pot boil and realized that the movement of the lid bouncing up and down implied that steam could be used to produce mechanical energy. This led to experimentation with steam, the invention of the steam engine, and the beginning of the industrial revolution.

It was research into the nature of gravity, electricity, and magnetism that inevitably led to discovery of related fields and forces that exist at the microscopic level of nature, of the short-range forces within atoms, for example. These phenomena, even if they are otherwise invisible to us, are related to our initial fields of inquiry. It is very likely that such relatedness is defined by our capacity to recognize the phenomena in the first place. This says something about the way our minds are capable of perceiving, or working with our sense perceptions. Lodestone and amber set the scene for a particular type of thinking, which gave birth to the scientific age. Without the magic of magnetism to inspire and stimulate the imagination, it is highly likely that the evolution of human thought would have been different.

Throughout our story of magnetism, we found that scientists often overlooked that which stared him in the face, at least as judged by hindsight. Thus a compass-maker who found his needles pointing downward added counterweights to balance them and for years missed the point. Similarly, Ampère missed the discovery of induction, and at least six physicists missed discovering radio waves because these various phenomena did not fit with their expectations.

The human mind weaves a tortuous path through the jungles of discovery, slowly wending its ways toward greater understanding, never really knowing what lies around the next bend, always astonished when a new vista is revealed, always ready and able to exploit what it has found for the benefit of its own species. So now we come to the larger picture. What about life? What about existence? Are we on the right trail? Where will our search lead?

What appears so awesome is that nature has provided us with the toys we needed to propel us into exploration of the inner secrets of the physical universe. But then our species is only what it is, and has only evolved as far as it has toward understanding the nature of our world because of the nature of that world. We are part and parcel of life on earth and it is not surprising that we have learned about magnetism and electromagnetism thanks to bacteria. It could not have been otherwise. After all, we are here because of the eternal interplay between hundreds of millions (perhaps billions) of species of which we are but one. Most of them have long been extinct, and many of them were and still are bacteria. We are part of a vast living heritage that extends back at least four billion years. We are the way we are only because of the way all life on earth has been, and is. If any single factor had been different, we would not be here to talk and think about it; at least not today!

This is the conclusion reached by Gould in reporting on the marvelous display of fossils associated with the Burgess shale beds in Canada. He wrote that "almost every interesting event in life's history falls into the realm of contingency."[28] Contingency concerns chance events that are not inherently predictable, which have determined the outcome of life's experiments. As Gould put it, if we replay the tape of life, even beginning from an identical starting point such as suggested in the Burgess shale, "the chance becomes vanishingly small that anything like human intelligence would grace the replay."[29] The replay might not include the GS-15 bacteria, hence no lodestone, hence no development of the science of electromagnetism such as we have recognized in history. Similarly, we recognize that very specific events played a profound role in the development of modern technology. We are here, living the life we live, only because those events, such as Hertz recognizing a tiny spark where he had not expected to see one, happened when they did. From tiny acts come enormous consequences, and it is fun to look at some of those acts from our perspective in time.

NOTES

1. In the words of one writer, the difference between the two types of explanation rested in how the force was treated mathematically: ordinary differential equations were used to describe action at a distance, as was true for the Newtonian treatment of gravity as well. Partial differential equations expressed the behavior of a field. A. E. Woodruff,

"Action at a Distance in Nineteenth Century Electrodynamics." *ISIS* 53 (1962): 439.

2. Thus gravitational force, F, was given by Newton's relationship

$$F = G \frac{m_1 m_2}{r^2}$$

where G is the gravitational constant and m_1 and m_2 are the two masses and r is the distance between them. Coulomb's law for the force E between electrical charges, e_1 and e_2, was

$$F = \frac{e_1 e_2}{r^2}.$$

Both equations involve distance squared in the denominator; neither equation involves time.

3. Paul Lagacker and Alfred K. Mann, "The Unification of Electromagnetism with the Weak Force." *Physics Today* (December 1989): 22.

4. Derek R. Lovley, John F. Stolz, Gordon L. Nord, Jr., and Elizabeth J. P. Phillips, "Anaerobic Production of Magnetite by a Dissimilatory Iron-Reducing Microorganism." *Nature* 350 (1987): 252.

5. N. Petersen, T. von Dobeneck, and H. Vali, "Fossil Bacterial Magnetite in Deep-Sea Sediments from the South Atlantic Ocean." *Nature* 320 (1986): 611.

6. R. Blakemore, "Magnetotactic Bacteria". *Science* 190 (1975): 377.

7. That magnetite existed in the cells of the magnetotactic bacteria was shown by R. Frankel, R. P. Blakemore, and R. S. Wolfe, *Science* 203 (1979): 1355.

8. B. Frankel, *Nature* 320 (1986): 575.

9. J. W. E. Fassbinder, H. Stanjek, and H. Vali, "Occurrence of Magnetic Bacteria in Soil." *Nature* 343 (1990): 161.

10. See, for example, M. M. Walker and M. E. Bitterman, *Journal of Experimental Biology* 145 (1989): 489.

11. For a review see J. L. Gould, "The Map Sense of Pigeons." *Nature* 296 (1982): 205.

12. See, for example, R. Beason and P. Semm, *Neuroscience Letters* 80 (1987): 229.

13. Many books, in particular proceedings of scholarly meetings, discuss these topics. For example, the reader is referred to *Animal Migration, Navigation and Homing,* edited by K. Schmidt-Koenig and W. T. Keeton (Berlin and Heidelberg: Springer-Verlag, 1978) and *Biophysical Effects of Steady Magnetic Fields,* edited by G. Maret, N. Boccara, and J. Kiepenheuer (Berlin and Heidelberg: Springer-Verlag, 1986).

14. M. M. Walker, M. E. Bitterman, and J. L. Kirschvink, "Exper-

imental and Correlational Studies of Responses to Magnetic Field Stimuli by Different Species." In *Biophysical Effects of Steady Magnetic Fields.*

15. Marc McCutcheon, *The Compass in Your Nose and Other Astonishing Facts About Humans.* (Los Angeles: Jeremy Tartcher, 1989).

16. Bern Dibner, *Oersted and the Discovery of Electromagnetism.* (New York: Blaisdell Publishing Company, 1962), p. 42.

17. L. Pearce Williams, "What Were Ampère's Earliest Discoveries in Electrodynamics?" *ISIS* 74 (1983): 492, specifically p. 505.

18. *Anales de Chemic* 30 (1801): 209, as reported by Dibner, *Oersted and the Discovery of Electromagnetism.*

19. While working on a revision of this chapter, I saw my first Aurora Borealis, which covered the entire sky visible out of an airplane window over the North Atlantic. I could not imagine how anyone seeing that phenomenon would infer the presence of an invisible field of force, magnetism, as essential for its explanation unless terrestrial magnetism was already known to exist. Even then it took a long time to figure out the explanation.

20. See Chapter 12 for a discussion of the Zeeman effect, which accounts for this phenomenon.

21. Motion of a source of light (or any electromagnetic radiation) relative to the observer produces a shift in the wavelength observed. This is known as the Doppler effect. The shift in wavelength depends on the velocity of the source. In its most popular guise, the Doppler effect produces the so-called *redshift* of light (shifted to longer wavelengths) from distant galaxies, which led astronomers to conclude that the universe is expanding.

22. See Chapter 12.

23. William Gilbert, *De Magnete* Book II, Chapter 11. Translation by P. Fleury Mottelay. (New York: Dover Publications, 1958).

24. Ibid.

25. Ibid.

26. Ibid.

27. Stephen Jay Gould, *Wonderful Life.* (New York: W. W. Norton, 1989).

28. Ibid.

29. Ibid.

⌐ 12 ⌐
Magnetic Fields in Space

Magnetic fields are to astrophysics what sex is to psycho-analysis.

H. C. van de Hulst

It is all very well to theorize, but it is what we learn from experiment that really counts.

Anonymous

T HAT mysterious and elusive property of matter known as magnetism is fundamental to our understanding of astronomical phenomena. Magnetism exists in planets and stars, galaxies and quasars. Magnetic fields permeate and power the shattered remnants of stars that died violently thousands of years ago. Magnetic fields thread their way through space between the stars in the Milky Way. How do astronomers know this? And why is magnetism barely discussed, even at scientific meetings?[1] The answer to the second question is that in astrophysics magnetic fields are difficult to measure and even more difficult to take into account in theoretical models. As Professor van de Hulst suggested, in most situations astronomers prefer not to talk about magnetism, yet, like sex, its pervasive influence is felt everywhere. The answer to the question about how astronomers detect magnetic fields in space involves another saga of discovery, the subject of this chapter.

Everything we know about the distant universe comes through a study of electromagnetic waves such as light, X-rays, radio waves, ultraviolet, and infrared radiation. Little of this knowledge would now be available if it were not for humankind's ability to build radio receivers, X-ray detectors, and bigger and better optical telescopes. All these things, need I remind the reader, are possible because of the understanding of electricity and magnetism that grew in the laboratories of the scientists already mentioned in this book. As to the discovery of magnetism in astronomical objects, we must again turn back the pages of history, this time to the year 1896, in a laboratory at the University of Leiden in the Netherlands.

A young man of thirty-two was about to perform a historic experiment. Pieter Zeeman (1865–1943), the son of a Dutch Lutheran minister, was warned as a student not to study physics. "Physics is no longer a promising subject," he was told, "it is finished, there is no room for anything really new."[2] Despite this caution he chose physics as his career and soon proved himself to be a capable and imaginative scientist. One day he considered repeating an experiment he had attempted unsuccessfully on a number of previous occasions. He had a far-fetched notion that magnetism and light were related but had not been able to prove it. His early attempts to show a connection had failed. Then he read, in a biography of Michael Faraday, that Faraday had believed in the connection between magnetism and light. This inspired Zeeman to try again. "If Faraday thought about [this] experiment," he reasoned, "perhaps it might yet be worthwhile to try the experiment again with the excellent auxiliaries of spectroscopy of the present time."[3]

Before he could begin he needed the approval of the laboratory director, Kammerlingh Onnes, who refused to give it. Onnes said that Zeeman had already shown the experiment wouldn't work, and he didn't want to hear any more about the subject. He considered the experiment a waste of time and resources, and he gave Zeeman instructions not to proceed. Disappointed but not disheartened, Zeeman shared the bad news with his friend and colleague, Hendrik Lorentz (1853–1928), a theoretical physicist, who agreed with Zeeman that the experiment might reveal something interesting.

Zeeman bided his time. When his boss left town on business he went into action. He placed a sodium flame, produced by burning common salt soaked in a piece of asbestos, between the poles of a powerful Rühmkorff electromagnet that produced a magnetic field of 10,000 gauss. When heated to incandescence, the sodium in the salt produced a vapor that glowed with a rich yellowish color due to two bright spectral lines, the sodium D-lines. He used a newly invented (Rowland) grating, a device that enabled him to see spectral lines more clearly. His hope was that when the magnet was switched on something would happen to the appearance of the spectrum.

He set up his equipment very carefully, despite being in somewhat of a hurry to complete the work before Onnes returned. Perhaps the ghost of Faraday hovered nearby to urge him on as the light from the flaring yellow flame shone brightly against a black background. Zeeman threw the switch and current surged into the giant coils (Fig. 12–1). The powerful 10,000 gauss field leaped between the poles of the magnet, and the sodium-D lines became about three or four times their original width!

Zeeman was ecstatic. Whatever the observation signified, it was proof that light and magnetism were related. That was enough to justify having done the experiment. However, when Kammerlingh Onnes returned and heard that Zeeman had disobeyed his orders he promptly fired him. It did not make for extenuating circumstances that Zeeman had succeeded. What was important was that orders had been disobeyed. The University of Amsterdam quickly offered Zeeman a job. He had the last laugh when in 1902 he shared the Nobel Prize in physics with Lorentz "in recognition of the extraordinary service they rendered by their research into the influence of magnetism upon radiation phenomena."[4]

When he published his results, Zeeman was cautious. "Possibly the observed phenomena will be regarded as nothing of any consequence," he wrote. Little did he know the far-reaching consequences his work would have. His discovery stimulated a flood of work on both the experimental verification and theoretical aspects of the phenomenon of spectral line splitting in the laboratory.

The Zeeman effect, as it would be called, was soon interpreted as proof that something was moving around inside atoms, something that could be influenced by the field. This turned out to be a charged particle—the electron—the same particle that carried current in conductors in experiments performed by

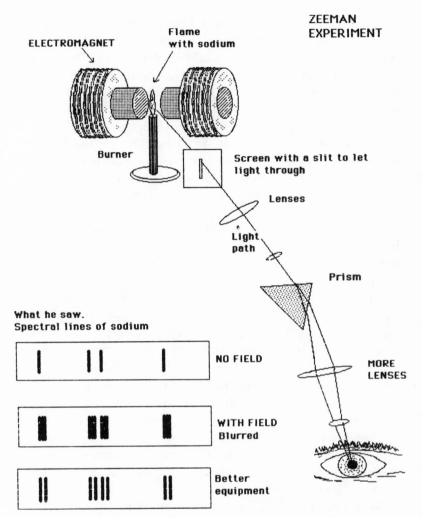

Figure 12–1. A schematic diagram of the experiment performed by Pieter Zeeman in 1896. The light from a sodium impregnated flame placed between the poles of a large electromagnet is passed through a slit, lenses, a prism, and more lenses so that Zeeman could see the spectral lines produced by the sodium. The inset shows what such lines might have looked like to him in the absence of a field. When the magnetic field was switched on, the lines broadened. Subsequently, using better apparatus, Zeeman found that the lines were split into several components.

Faraday and Hertz, even if this was only just beginning to be recognized. Later it was realized that these particles, in motion, had to be responsible for the spectral lines themselves.

Zeeman showed that each of the sodium D-lines was not really broadened but was split into several distinctly different lines where before there had been only one. He went on to show that in the presence of a magnetic field the sodium D-line was split into ten components. The splitting allowed the properties of the electron to be measured, something that was worked out by Lorenz, in particular the ratio of the charge of the electron to its mass, a very important quantity in physics.

Zeeman also found that the multiple components of the sodium line each had its own polarization, which depended on how one viewed it relative to the field. Here we must digress briefly. In general, the study of magnetic fields in the universe involves the observation of *polarization*. To illustrate the phenomenon, consider an analogy. Radiation traveling through space is like a wave running along a stretched rope. If the rope is flicked up and down at one end, a *linearly polarized* wave is created and its *plane* of polarization is said to be vertical. A sideways flick generates a horizontally polarized wave. A third alternative is to rotate one end of the rope to generate a *circularly* polarized wave. Astronomers can use filters to measure the state of polarization of light, for example, by using what are essentially polaroid glasses and rotating them in front of the photographic plate and noting at what angle a particular star appears brightest. This gives the direction of polarization for that star.

In the case of radio waves, polarization can be measured by rotating the antenna around an axis aligned with the direction from which the radio waves are coming. If you have ever tried to adjust an indoor TV antenna you may have noticed that different channels require the antenna to be oriented either horizontally or vertically. You need to match the polarization of the transmitter to maximize the received signal and improve your reception. It is so for radio waves from space as well.

Zeeman's discovery opened the way for future astronomers to find magnetic splitting of spectral lines in astrophysical situations. The polarization of the spectral lines that were shifted in wavelength in the presence of a field gave the clue that would allow splitting to be positively identified. In 1908 George Ellery

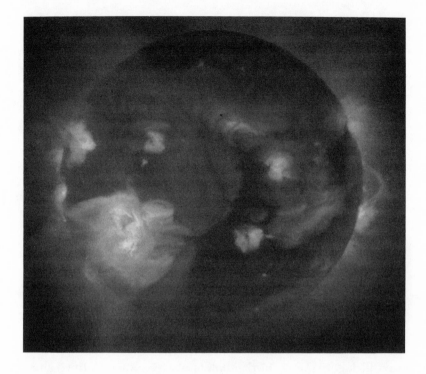

Hale, at Yerkes Observatory, discovered the Zeeman effect in the light from sunspots and inferred field strengths of thousands of gauss. He, too, was surprised by what he had found and entitled his report "On the probable existence of a magnetic field in sun-spots."

Today there are a dozen telescopes in the world that do nothing else but measure the Zeeman splitting of the light from the sun in order to map magnetic fields in and around sunspots (Fig. 12–2). These reveal that the sun's basic field is about 1 gauss strong, runs north–south, and reverses its direction every twenty-two years. Within sunspots the field reaches about 1000 gauss, and pairs of spots appear to act as the north and south poles of huge magnets many hundreds of times the size of the earth. The fields in pairs of sunspots also change their relative polarity with the eleven-year sunspot cycle.

Thirty-eight years passed and then, in 1946, Horace Babcock at Mt. Wilson Observatory observed the Zeeman effect in the light from another star, 78 Virginis. This was the first ob-

Figure 12–2. This X-ray image of the sun vividly highlights magnetic fields that surge out of the solar surface and reach way up into the corona, the sun's atmosphere. Gas in the corona is normally at a temperature of a million degrees and in active regions, the bright patches in this image, the temperature rises to three million degrees, causing the emission of powerful X-rays. Hot gas flowing along the magnetic field lines outlines their shape and in places the patterns remind us of Faraday's sketches. At the edge of the sun, prominences rise ten thousand kilometers above the surface. This image was acquired by the Soft X-ray Telescope on the Yohkoh solar research spacecraft of the Japanese Institute of Space and Astronautical Science. The Soft X-ray Telescope experiment is a Japan/U.S. collaboration involving the National Astronomical Observatory of Japan, the University of Tokyo, and the Lockheed Palo Alto Research Laboratory. The U.S. work is supported by the National Aeronautics and Space Administration.

For the technically minded, the intensity of this image is displayed by a logarithmic scale and the intensity range is about 140,000. There is no flare in progress at the time or the intensity range would have easily exceeded a million. This dynamic range is unprecedented and is one of the strengths of the Soft X-ray Telescope. The sunspot region which produced all of the major flares over the period surrounding this picture is the complicated bright structure in the lower left (SE quadrant of the solar disk). North is at the top and east is to the left. This picture was taken on October 25, 1991.

servation of a number of the so-called magnetic stars whose fields range from 100 to 34,000 gauss.

The first direct evidence for a magnetic field between the stars was actually stumbled on by accident in 1949 when two astronomers, W. A. Hiltner and J. S. Hall, were attempting to measure polarization of starlight by observing eclipsing binary stars.[5] They found the binary stars to be unpolarized, which destroyed the theory they had been testing! However, other stars they had observed to test their equipment was working—comparison stars—were polarized, and their amount of polarization was correlated with the amount of interstellar dust through which the stars were shining.

Theories were then developed that invoked the existence of

elongated dust particles between the stars. A magnetic field influences the motion of slightly elongated particles so they tend to line up. Starlight then interacts with these aligned particles so that light that passes through a cloud of dust emerges with a small net polarization parallel to the field direction.

Hall and Hiltner's observations represented the first evidence for the existence of magnetic fields in the galaxy. Optical polarization measurements of thousands of stars have since been made, and when these data are plotted on a map of the sky they reveal the overall pattern of magnetic fields in interstellar space. In general, the field follows the band of the Milky Way, which we can see in the night sky (especially in summer in the Northern Hemisphere), but it also exhibits distinct anomalies that may be associated with the presence of remains of stars that exploded millions of years ago to blow vast magnetic bubbles in interstellar space.

Models for the interaction between interstellar dust grains and magnetic fields to account for optical polarization required field strengths of 10 microgauss.[6] However, the field strength has to be independently determined to confirm the model.

Once again the relevance of Zeeman's discovery presented itself. In 1959 John Bolton and Paul Wild, two Australian radio astonomers, suggested that it should be possible to measure the Zeeman splitting of the spectral line at the 21-cm radio wavelength produced by hydrogen gas between the stars. They realized that for a field of a few millionths of a gauss (microgauss) the amount of splitting was going to be very small, about one-thousandth of the width of the spectral line in question. It would be impossible to measure this directly, but there was a way to detect it if two circularly polarized antennas were used—one sensitive to radio waves rotating clockwise, the other to counterclockwise polarization. Comparison of what was observed with each polarization simultaneously would allow the Zeeman effect to be detected and hence the field strength in the gas cloud could be estimated.

This suggestion set off a chain of experimentation at several observatories. In 1961 I joined a team of radio astronomers at the Nuffield Radio Astronomy Laboratories at Jodrell Bank in England and spent six fruitless years trying to detect the Zeeman effect in the interstellar hydrogen clouds. In retrospect, our equipment was not up to the task. The signals we were hunting for were so weak that spurious polarizations caused by imperfect telescopes and antennas ruined our chances of success. At Jodrell Bank we used over two thousand hours of observations in a vain search. As much time again was fruitlessly used at other radio telescopes around the world in this elusive quest.

Finally, in 1968, using data obtained with the 140-foot radio telescope of the National Radio Astronomy Observatory in Green Bank, West Virginia, I found the first evidence for the Zeeman effect lurking in the circularly polarized radio signature from hydrogen clouds. The search uncovered fields of 10 and 20 microgauss, 100,000 times weaker than the earth's field, in clouds seen toward the radio-emitting remains of a 200-year-old supernova in the direction of the constellation Cassiopeia. To me it still seems remarkable that it is possible to detect a field so weak in a region of space 10,000 light years from earth, and all because Pieter Zeeman had surreptitiously performed his experiment back in 1896, an experiment he later thought would be of little interest to anyone else.

Today the search of Zeeman effects at 21-cm continues at the Hat Creek Radio Observatory of the University of California in Berkeley, where Carl Heiles and his colleagues have amassed thousands of hours of observations to detect fields in the Milky Way, and at the National Radio Astronomy Observatory in Green Bank, where I map the magnetic field patterns in small areas of sky. Interstellar fields of a few microgauss are now routinely detected. Apparently this is the strength of the average field between the stars. It is typically up to 10 microgauss in vast filamentary tentacles of gas and dust that weave their way throughout the galaxy. These fields are probably just strong enough to account for the alignment of interstellar dust grains.

Interstellar space[7] is also laced with free electrons, the carriers of current electricity. These electrons have been driven out of their parent atoms by the diffuse glow of starlight, or by the shattering impacts of shock waves created when stars disintegrate. The atoms are said to be ionized in the process. The ionized atoms and the electrons move through space, and here we run up against a fascinating aspect of the story. Faraday had shown that moving magnetic fields drive currents, and Oersted discovered that currents (moving electrons) create magnetic fields. Since interstellar space is filled with electrically charged matter in constant motion, wouldn't we expect those currents to create fields? Indeed, they do. But if the fields also move about, would that not create new currents, which would create fields, etc., etc., ad infinitum? This endless loop may hold the key to explaining magnetism in and between the stars. Once a field is created in an ionized medium, such as in the core of a star, or in space between the stars, the field maintains itself because of this endless interplay between current and field. Provided no energy is lost, this dance can continue indefinately. Within the gas clouds in space, magnetic fields become "frozen-in"; that is, they are part of the cloud. If the gas moves, the field moves with it, and vice versa. If the cloud contracts, the fields will become stronger and can actually stop the contraction because fields do not like to be crowded. Understanding this interplay between gas and fields forms a science called magneto-hydro-dynamics.

Closer to home, frozen-in fields in the sun bubble to the surface in and around sunspots where small blobs of gas are driven upward. The interaction between segments of frozen-in field can cause violent solar explosions known as flares. The role of magnetic fields in the sun is so important that Robert Leighton, an astronomer at Caltech, is quoted as saying, "If the sun had no magnetic field, it would be as uninteresting as most astronomers think it is."[8]

Another clue concerning the nature of fields between the stars first emerged in the 1960s. Radio waves from the Milky Way and from extragalactic radio sources were found to be linearly polarized at levels of a few percent. The angle of polarization is different at different radio wavelengths as the result of an analog of a remarkable phenomenon discovered by Michael Faraday in 1845 when he shone polarized light through a certain type of glass permeated by a magnetic field.

In astronomical situations the amount of Faraday rotation—that is, the rotation of the plane of polarization while the wave travels through space much as a corkscrew rotates—allows two important things to be derived. First, the polarization angle at the location where the radio waves originated can be found. In the case of distant galaxies, this allows the magnetic field patterns in those galaxies to be mapped. In the case of the Milky Way it allows the field structure in the region where the radio waves originate to be visualized.

Second, the direction of rotation (clockwise or counterclockwise as wavelength changes) reveals the average direction of field between the source and the sun; that is, whether it is directed toward or away from us. If assumptions are made about the depth of the clouds of electrons responsible for the rotation, it is also possible to estimate the field strength within them.

Polarization measurements of radio waves from the Milky Way have involved many Ph. D. students during the last thirty years, in particular in the Netherlands and at Cambridge University and Jodrell Bank in Great Britain. Titus Spoelstra at the Dwingeloo Radio Observatory combined all the Dutch data on the polarization of radio waves for the Milky Way and concluded that the radio signals involved originate at a distance of up to a thousand light years.

Masato Tsuboi, Makato Inoue, and other radio astronomers at Nobeyama Radio observatory in Japan have made a speciality of studying radio polarization near the center of our Milky Way galaxy and found what appears to be a magnetic jet emerging from the center. Yoshiaki Sofue, of the Institute of Astronomy at Tokyo University, is one of a growing band of researchers who believe that this is evidence for a primeval origin of the magnetic field. The ancient field was wound up as the galaxy rotated and is now concentrated near the galactic center where it is tightly wound. "This might be a magnetic tornado," he says, conjuring up a poetic image of an enormous swirl of magnetic fields in the very heart of the galaxy.

Observations of Faraday rotation of hundreds of distant radio sources allow the field patterns in those radio sources as wells as the mean field direction on a scale comparable with the depth of our galaxy in the direction of the sources to be found. Philipp Kronberg and his collaborators at the University of Toronto have interpreted such data to draw a picture of the large-scale magnetic field in the Milky Way. Close to the sun it follows the distribution of stars in what is known as the local (or Orion) spiral arm and beyond a few thousand light years distance it seems to encircle the center of the Milky Way.

Further clues to help us understand magnetism in the Milky Way come from a study of magnetic fields in other galaxies. Richard Wielebinski, Marita Krause, and Rainer Beck, at the Max Planck Institute for Radio Astronomy in Bonn, have been mapping fields in nearby galaxies by combining radio and optical observations. As is often the case in science, what they have found may not help clarify the picture as much as it shows that the universe is quite capable of being perverse in its action. Some galaxies, such as M 51, show fields directed along the spiral arms (Fig. 12–3). Several galaxies have fields that reverse their direction from spiral arm to spiral arm. This means the fields run into as well as out of their centers. They are known as bi-symmetrical spirals.

Some galaxies show fields either all going into or all coming out of their centers where the fields must either enter from, or emerge into, intergalactic space. One edge-on spiral, NGC 4631, shows magnetic fields radiating outward above and below the disk. Another odd case, NGC 4258, was observed by G. D. van Albada, J. M. van der Hulst, and J. H. Oort in the Netherlands. This galaxy shows magnetic spiral structure in perfect anticoincidence with the optical spiral arms.

All this suggests that galaxies may be as unique in their magnetic configurations as they are in shape. Meanwhile, we are stuck inside the Milky Way trying to figure out what the magnetic patterns in our galaxy are like. Unfortunately, earth is not an ideal vantage point from which to observe the truth about the Milky Way, especially its large-scale properties. But for the moment this location must do.

The total amount of Faraday rotation found in the direction of a given source of radio waves depends on the field strength throughout the clouds of electrons that lie in intervening space. Using the distance through the intervening clouds of known particle density the field strength can be derived. But since neither the depth nor the density of the interstellar clouds can be accurately estimated, there appears to be no way to unambiguously derive the field strength. This was the situation until pulsars were discovered.

Figure 12–3. The magnetic field pattern in the spiral galazy M51 superimposed on an optical image of the galaxy. The magnetic fields clearly follow the spiral arms in this galaxy. Courtesy N. Neininger, R. Beck, U. Klein, and R. Wielebinski, Max-Planck Institut für Radioastronomie, Bonn, Germany.

Pulsars are peculiar objects, neutron stars that are the remains of cataclysmic stellar explosions called supernovae. Pulsars transmit a stream of incredibly regular pulses of radio energy, which astronomers on earth have been studying for over two decades. Those pulses provide a marvelous tool for studying magnetic fields in space, fields that have absolutely nothing to do with the pulsars themselves. In fact, the technique would work even if we had no idea of what a pulsar is.

The radio pulses from a pulsar are linearly polarized. Neutron star theories invoked to account for pulsar polarization require fields in those amazing neutron stars to be a staggering trillion (10^{12}) gauss. The escaping radio waves are soon free of the pulsar and, on the way to earth, pass through interstellar clouds of ionized material that cause the plane of polarization to suffer Faraday rotation. This rotation has been measured for some 160 pulsars by Andrew Lyne at Jodrell Bank as well as by other radio astronomers. The trick to using the pulsar data is elegant in its simplicity. A radio pulse experiences an effect known as dispersion, a measure of the delay in arrival time of the pulse as a function of wavelength. The amount of dispersion depends on the total number of electrons between the pulsar and the telescope. This happens to be just that quantity required to interpret Faraday rotation data. If the total number of electrons between the source and the sun is known, from the dispersion measure, the average magnetic field strength in the gas clouds producing the rotation can be found unequivocally by simple arithmetic, dividing the rotation measure by the dispersion measure.

The two observed quantities, rotation measure and dispersion measure, are on record for several hundred pulsars and have allowed the intervening field strength to be estimated. Richard Rand and Shrinivas Kulkarni at Caltech used all available data, including those for forty or so pulsars observed by Richard Manchester of the Commonwealth Scientific and Industrial Research Organization in Australia, to conclude that the mean magnetic field within several thousand light years of the sun is about 2 microgauss and is concentric about the center of the galaxy, a result consistent with Kronberg's conclusions.

Despite the obvious existence of magnetism in the Milky Way and in other galaxies, the origin of the fields remains a mystery. Here is a classic case of deciding which came first. Was it the chicken, the magnetic field, or the egg, electric currents? One school of thought invokes a dynamo that involves motion on a galactic scale. This would create an underlying field then pushed hither and thither and amplified by localized events such as supernova explosions. Another theory assumes that the primeval magnetic field came first and was wound up as the galaxy formed. This "seed" field, a remnant of the Big Bang, may later have become amplified within the galaxy by a dynamo effect such as occurs inside the earth where swirling eddies of liquid rock generate magnetic fields.

The pervasive force of magnetism in space influences physical processes on all scales from star formation to the evolution of the galaxy. In learning how to study magnetic fields in the Milky Way, astronomers have come a long way from those superstitious days of yore when a lodestone was regarded with mystical awe. Now the magnetic field structures of the Milky Way, and in galaxies millions of light years away, are grist for the mill of science.

NOTES

1. An official at one of the nation's primary sources of funding for astronomical research told me recently that if I wrote a grant proposal I should try to avoid reference to magnetic fields (in interstellar space in this case) because the referees had a bad habit of rejecting any proposal that deigned to mention the topic.

2. Hendrik Casimer, *Haphazard Reality*. (New York: Harper & Row, 1983), p. 28.

3. A. Pais, *Inward Bound*. (Oxford: Oxford University Press, 1986), p. 76.

4. Ibid., p. 78.

5. Binary stars are pairs of stars that orbit one another. When their orbits are suitably aligned relative to the earth, each star may pass behind the other in a cyclical manner over a period of years. Each time one of the stars is in front of the other it creates an eclipse.

6. A microgauss is one-millionth of a gauss.

7. See Gerrit L. Verschuur, *Interstellar Matters*. (New York: Springer-Verlag, 1989) for a full story of what exists between the stars.

8. Quoted by Jeffrey L. Linsky, *Solar Physics* 121 (1989): 187.

n 13 n
The Spark That Bridged the Universe

The Catholic Church . . . seized on the big bang model and in 1951 officially pronounced it to be in accordance with the Bible.

Stephen Hawking, *A Brief History of Time*

I N dingy and dark laboratories in the nineteenth century, discoveries were made by Faraday and Zeeman that someday would allow astronomers to probe into the nature of magnetism in stars, between the stars, and in distant galaxies. And the spark that Hertz discovered in his happy experiment would someday allow astronomers to bridge the universe, to peer back to the very beginning of time. Hertz's "electric waves" are now used to study the faint glow that is an echo of creation, a whisper left over from the cataclysm that precipitated the cosmos into existence: the Big Bang!

Many key discoveries in science were at first missed because the person first presented with the phenomenon wasn't looking for it. Thus Ampère missed discovering induction. But Hertz found fame by pursuing an unexpected new phenomenon. His mind was willing to comprehend the stray side spark and thus began his two-year series of experiments that changed the course of human history.

In his experimentation Hertz demonstrated that radio waves could carry energy over more than twenty meters, which was about as far as he could move in the lecture theater that was his laboratory. Since he was interested in other aspects of his research, he did not ask how much further they might travel. That was up to the Italian physicist Guglielmo Marconi, who, upon hearing of Hertz's discovery, asked, "How far can Hertzian waves travel?"

In 1901 Marconi found a practical answer when he transmitted Morse code for the letter S across the Atlantic Ocean. That was the beginning of long-distance radio communication. It was a decade before vacuum tubes were invented; by the late 1950s transistors burst upon the scene and triggered the transformation of radio technology into the marvel it is today.

Most of us take for granted that radio waves travel around the world and across the depth of the solar system. After all, NASA engineers have little trouble communicating with the Voyager spacecraft ranging well beyond Neptune's orbit. Hertz and Marconi would have been stunned to realize that radio waves can travel so far and still be detectable. They would have been even more overwhelmed had they lived to see the result of an experiment carried out in 1933. In that year Karl Jansky at the Bell Telephone Laboratories was asked to investigate why there was so much unwanted noise (crackles and hisses) on transatlantic telephone links. He discovered three sources: near lightning and distant lightning, and a steady hiss from the Milky Way! This marked the birth of radio astronomy, the study of "Hertzian" waves arriving from the depths of space. Now the answer to Marconi's question took on a new dimension. It was possible to detect radio waves that had traveled from the center of the Milky Way, about thirty thousand light years distant.

By the 1960s radio astronomers discovered radio waves arriving from distant galaxies. Hertz's little spark had opened a Pandora's box of discovery. A hop across the Atlantic was less than insignificant compared with journeys across hundreds of millions of light years from those galaxies.

In 1963 quasars, sources of strong radio emission located beyond most visible galaxies, were first observed. Radio waves from certain quasars have traveled for billions of years. It was not long before astronomers started to talk about "seeing" so far back in time that they might be able to detect radio waves that began their journey to earth soon after the universe was created in the so-called Big Bang. In other words, they imagined receiving radio waves that have traveled since the beginning of time and across the entire observable universe. Astronomers discussed this possibility even as engineers at Bell Telephone Laboratories inadvertently found evidence for such radio signals, without realizing they had done so.

Here our tale again becomes one of lost opportunities. In a sequence of experiments that began in 1959, R. W. de Grasse, D. C. Hogg, E. A. Ohm, and H. E. D. Scovil at Bell Labs experimented with a radio antenna specially designed to produce as little electrical "noise" as possible.[1] Such noise hampered the prime purpose of the antenna: to pick up weak satellite broadcasts. They reported that the amount of noise received while the antenna pointed at "empty" sky was equivalent to 18 degrees Kelvin (K),[2] of which they could account for 16 K from a variety of sources, such as stray radiation from the ground and the atmosphere, and noise produced by the radio receiver and the antenna. This left 2 K unaccounted for. The cause of this discrepancy was not specifically addressed in their report.

Within a year Edward Ohm published measurements made with another receiver connected to the same antenna and found he could account for only 19 K of the 22 K signal that was measured.[3] Although the difference of 3 K was again not explicitly referred to, the number looms clearly in a table of data in his report.

In 1963, using the equipment with which the first transatlantic satellite signals were broadcast by "Telstar," Bill Jakes, also of Bell Labs, reported on a similar set of antenna measurements.[4] He reported about 2.5 K more than could be accounted for by known sources of electrical noise.

With hindsight it is clear that these researchers had all detected the radio signal that would later be more carefully studied and earn Arno Penzias and Robert Wilson, also of Bell Telephone Laboratories, a Nobel Prize in physics. They would take their place in history because they did what Hertz had done when he first spotted his "side spark." They asked "Why?" Their quest for an answer benefited from a fortuitous telephone conversation.

Penzias and Wilson, using the same antenna as the others, also found the 3 K signal. But they happened to talk to astronomical colleagues about it. One of those was Bernard Burke of MIT, who told them that radio astronomers at Princeton were searching for just such a signal. Cosmologists considering the nature of the Big Bang had predicted that the universe should be basking in a faint glow. This remnant of the cosmic fireball might be observable in the form of a radio signal that should be arriving from all directions, should be uniform over the sky, unpolarized and unvarying with time.

Penzias and Wilson's next task was clear. They would concentrate on finding whether the 2.7 K (its present value) signal had these properties. It did, and they entitled their report "A Measurement of Excess Antenna Temperature at 4080 Mc/s."[5] No longer was the excess signal left essentially unnoticed in a table. This time it was the focus of their work. In the same issue of the journal, R. H. Dicke, P. J. E. Peebles, P. G. Roll, and D. T. Wilkinson of Princeton explained why such a signal was expected from the Big Bang, the event that created our universe and set space expanding.[6] The 2.7 K signal was also given its own name: the cosmic microwave background (CMB).

Now the answer to Marconi's question took on an amazing new significance. Radio waves that had traveled since the beginning of time had been detected on earth. The waves permeate the air all around us, and all we have to do to detect them is build a sensitive enough antenna and receiver system. But what was the nature of this glow, and where was it coming from?

Astronomy is a form of archeology. Distant objects are seen by the light (and other electromagnetic waves) they emitted a long time ago. For example, the sun is seen as it was eight minutes ago. In astronomical parlance, the sun's distance is eight light minutes. The nearest star beyond the sun is about four light years away, the Milky Way is about 100,000 light years across, and other galaxies are from millions to billions of light years distant.

Archaelogists would love to be able to do what astronomers can do—literally see the subjects of their study living in the past. By examining the relics of light, radio waves or X-rays that reach us today, astronomers can directly observe what distant galaxies were like a long time ago. Most astronomers deal with relatively nearby galaxies and even more neighborly stars. Therefore it is customary to ignore the fact that radiations from astronomical objects reach us long after they set course for earth. After all, what is a few hundred million years between friends, especially if the universe is about fifteen billion years old? In most astronomical situations no one pays much attention to the fact that light is "old" by the time it reaches us, except when it comes time to think about the beginning of the universe, the CMB, and the formation of galaxies.

Cosmic background radiation is seen arriving from all directions, but where is it really coming from? What are we actually "seeing" when we observe the 2.7 K signal? Consider how we "see" the sun. A typical photograph of the sun shows a perfectly round disk mottled by dark regions known as sunspots. At the edge of the disk, a fiery prominence or two might project up into the hot solar atmosphere called the corona (see Fig. 12–2). But why does the sun look like a sphere with a certain size if its hot gases actually extend well above the level we can see?

The answer provides us with a link to understanding the origin of the CMB. The visible disk of the sun, called the photosphere, is defined by a layer of gas whose temperature and density are such that light radiation can escape from the gas giving rise to that light. Light originating below the photosphere finds itself in a region where the density is so high that as soon as a photon is created it is immediately absorbed by matter around it. Above the photosphere conditions are different and light escapes freely between the particles of hot gas. In other words, immediately above the photosphere the sun's gases are transparent; below it they are opaque. When we look at a photograph of the sun we are seeing the photosphere, the layer where the sun becomes opaque or transparent, depending on how you look at it, from above or below. If we tried to shine a flashlight into the sun (an extreme case of tilting against windmills), the beam would get as far as the photosphere and then be absorbed.

The 2.7 K glow of the cosmic oven has traveled a very long time before it arrives at earth. It began its journey about fifteen billion years ago, soon after the universe was formed in the Big Bang. The radiation was originally created in the cosmic photosphere, a name given to an epoch about 300,000 years to one million years after the explosion, when the universe had cooled to about 3000 K, similar to the temperature of the surface of the sun (5800 K). At that time, the entire universe had a temperature and density similar to the solar photosphere today. Before then, photons of light created in the universe was absorbed by surrounding matter as soon as they came into existence. As the universe continued to expand and cool, light found itself free to escape the confines of the matter in which it was created. The universe became transparent, and light shone everywhere.

That light is said to have originated in the cosmic photosphere and began to travel in all directions, through all of time, until today, on earth, we see it arriving from all directions around us. The 2.7 K signal comes from all directions because when we look out into space, back into time, we see the universe as it was when it began as a hot fireball, provided we look far enough. Upon closer consideration this may appear very confusing. The Big Bang began at a point in space–time. Yet, when we look out into space, in all directions, we expect to be seeing a larger and larger volume defined by our ever-increasing horizon. But if we go far enough back in time (out into space), this large surface was actually smaller and smaller because we get closer and closer to the beginning of the universe. Therefore, we have to recognize that at great distances we are surrounded by what was once a point! The apparent paradox is dealt with by recognizing that space–time is curved; a subject well beyond the realm of our discussion.

By the time the light generated soon after the Big Bang reaches us it is extremely faint and is no longer in the form of light. The expansion of the universe has stretched the light waves (they have been redshifted) to the point where they reach us as longer-wavelength radio and infrared waves. The CMB carries more energy in the infrared region of the spectrum than in the radio, but the atmosphere absorbs most of the infrared before it reaches the ground. Therefore astronomers place infrared detectors on board satellites to get above the atmosphere in order to observe the CMB in the infrared.

If the universe were not expanding, the light produced in the cosmic photosphere would not be redshifted and the sky would be ablaze with light. In that case there would be no life on earth (it would be far too hot) and probably no stars or planets either. Stars would not form through the gravitational contraction of cool interstellar gas and dust clouds if all of space were very hot to begin with.

The answer to Marconi's question about how far Hertzian waves could travel is quite extraordinary. The cosmic microwave background radiation has traveled through all of time and space. What we now receive as radio waves began their journey as light! The waves have traveled since the beginning of time, or at least from so close to the beginning that it makes little difference as seen from a cosmic perspective. The CMB defines how far we can ever "see" into space. At this point some readers may recall reading or hearing about new telescopes that are claimed to allow us to see farther into space than others that have gone before. Those reports were nonsense! New optical telescopes may see "farther" in the visible light part of the spectrum than before, but they will never see as far as radio and infrared telescopes have already done. Radio telescopes, in particular, have been able to observe to the "edge" of space-time, the beginning of the universe, since 1959, even if the phenomenon was only recognized in 1965. Had the mirror of the Hubble Space Telescope been perfect it would not have been able to see farther into space than observers of the CMB have already done. This is because there is nothing to "see" beyond the cosmic photosphere (and it surrounds us), just as we cannot see into the sun through its photosphere (unless we use particles called neutrinos, which is another story).[7]

For about a century we have been certain that we live in a universe populated by galaxies. Once upon a time these galaxies must have been formed from vast clouds of matter that were created in the Big Bang. Therefore, if you want to study the clouds of matter out of which galaxies formed, all you need do is look far enough into space, back in time, to an epoch before galaxy formation occurred, and see what can be seen.

To look back in time is to do historical research. In principle it should be possible to study the epoch of galaxy formation by looking back to "somewhen" between the time when the first galaxies are known to have formed (at least thirteen billion years ago) and to the time when the cosmic microwave background was created when no galaxies existed (about fifteen billion years ago, give or take a few billion). This challenge has fascinated astronomers for decades, and they have long realized that the only way to explore galaxy formation was to measure the CMB carefully enough to find whether it has ripples in it. Such ripples or irregularities might signify the seeds of the emergence of protogalaxies, or protoclusters of galaxies. However, given that the signal we call the CMB is only 2.7 K to start with, the detection of tiny irregularities that would later evolve into swarms of galaxies was a tremendous challenge.

To measure the temperature of the CMB accurately, the Cosmic Background Explorer (COBE) spacecraft was launched in November, 1989. On board were sensitive thermometers that measured the temperature all over the sky to determine whether there were tiny fluctuations in its value.

Ground-based searches for inhomogeneities in the CMB began soon after its discovery, and with each more sensitive observation the CMB appeared to be smooth, although it did show a simply asymmetry. On one side of the sky the CMB is slightly cooler (redder) than on the other, owing to the motion of the sun and the Milky Way with respect to distant galaxies. Once this asymmetry was removed, the CMB was found to be smooth down to a hundredth of a percent or less.

With each new measurement of the isotropy (smoothness) of the CMB, theory had to be adjusted. Recently, theoretical explanations reached a crossroad, and so did the observations. There was only so much that could be done from the ground. To make more sensitive observations of the CMB, astronomers had to get above the earth's atmosphere to study the infrared and very short radio wavelength components of the CMB to more precisely define the nature of the radiation and measure its smoothness.

Within a few months of the launch of COBE, the principal investigators, John Mather, Michael Hauser, and George Smoot, announced the first results. The temperature of the CMB, the cosmic oven, is 2.735 K and its spectrum has a shape in perfect accord with the theory of how the radiation was produced in the cosmic photosphere. At the time this is being written, the COBE data show the CMB to be smooth down to levels of a few ten-thousandths of a percent, which poses enormous problems for theories of galaxy formation in a Big Bang universe.[8]

No one yet knows the answers to the mysteries of galaxy formation. There is an alternative school of thought led by Halton Arp, Sir Fred Hoyle, and Geoffrey Burbidge,[9] who have suggested that the CMB is not what we think it is and that the universe did not begin with a Big Bang. Theirs is a highly controversial point of view, which involves returning to a model in which the universe is in a steady state. According to them, the smoothness of the CMB is telling us that it originated very close to the sun and by a totally different process.

In the tradition of Hertz, Marconi, and Penzias and Wilson, the quest for answers to our deepest questions will be successful if we pay heed to the phenomena that nature lays before us. We can only become prepared to recognize answers after we have asked the right questions, and, as Pasteur said, we are not likely to find the answers by chance, unless our minds are prepared.

The tiny spark that flickered in Hertz's laboratory a century ago not only changed our lives by triggering the development of radio, radar, and television but opened the way for human beings to bridge the universe. Now we wonder if our notions about the universe may yet be changed again, perhaps by some metaphorical flicker even now being noticed (or ignored) by someone working in a laboratory, or with data from a spacecraft in orbit about the earth whose signals are being carried down to us on the back of Hertzian radio waves.

NOTES

1. R. W. de Grasse, D. C. Hogg, E. A. Ohm, and H. E. D. Scovil, "Ultra-low Noise Receiving System for Satellite or Space Communication." *Bell Telephone System Monograph* No. 3824, 1959.

2. Degrees Kelvin are the units on the absolute temperature scale used by many scientists, in particular astronomers. The zero point, 0 K, is equivalent to $-273°$C. If the strength of a signal observed with a radio telescope is said to be 18 K, it means the temperature of the radiating source appears to be $18°$ above absolute zero. It is a considerable technological challenge to measure accurately the temperatures (brightness) of sources of radio and other electromagnetic waves (such as infrared).

3. In *The Bell System Technical Journal,* July 1961.

4. In *The Bell System Technical Journal,* July 1963.

5. A. A. Penzias and R. W. Wilson, "A Measurement of Excess Antenna Temperature at 4080 Mc/s." *Astrophysical Journal* 142 (1985): 419.

6. R. H. Dicke, P. J. E. Peebles, P. G. Roll, and D. T. Wilkinson, "Cosmic Black-Body Radiation." *Astrophysical Journal* 142 (1985): 414.

7. Neutrinos are massless particles that travel near the speed of light. They barely interact with matter. In principle they can travel through the entire sun without being stopped. They are supposed to have been created in the very early universe and could have escaped at a time well before the cosmic photosphere became transparent to radiation.

8. The discovery of possible "ripples" in the cosmic microwave background was announced in 1992, and despite media enthusiasm it remains to be seen whether the problem of galaxy formation can be dealt with.

9. H. C. Arp, G. Burbidge, F. Hoyle, J. V. Narliker, and N. C. Wickramasinghe, "The Extragalactic Universe: An Alternative View." *Nature* 346 (1990): 807.

∩ 14 ∩
The Era of Creativity

Ask and it shall be given you, seek and you shall find.
Matthew 7:7

FOUR centuries after Gilbert we stand at an awesome time in the history of humankind. Four hundred years of increasingly persistent questioning of nature has brought us to a remarkable threshold. In the words of Stephen Hawking, there are "grounds for cautious optimism that we may now be very near the end of the search for the ultimate laws of nature."[1] However unlikely or even arrogant this claim may seem to some readers, it may contain more than a germ of truth.

In 1600 Gilbert cleared the decks of superstition surrounding lodestone and wondered how something could act like a magnet. Three centuries later, in 1900, after dozens of major scientists and hundreds of minor ones devoted their lives to exploring the nature of electricity and magnetism, the first comprehensive picture emerged of what the force of magnetism—as embodied in the theories that linked it to electricity—represented. But the questioning did not stop. There were things that still did not fit. Above all, scientists continued to ask the question, How do the different forces of nature tie together?

Significant progress in understanding the existence of matter and the forces that act on various forms of matter has occurred. For the first time it appears that humankind is on the brink of understanding everything.[2] Skeptics react to this statement and suggest that people have always thought this. But the issue is not whether a complete and self-consistent picture, the Theory of Everything, will emerge in the 1990s, or even in the early part of the twenty-first century. The point is that an end may be in sight. This does not imply that our species will learn all there is to know; instead, it means that the broad picture of the underlying forces and processes that play themselves out in the physical universe will be recognized. At no previous time in history could this have been claimed with as much conviction based upon the results of experiment and on available knowledge. Although Faraday, Zeeman, Einstein, and many other physicists were always seeking to unify their understanding of the various forces of nature, none of them succeeded. With the marriage of electromagnetism and the weak force, the first major step to a unification of all four forces was taken. Final success may be only a matter of time. The path is already visible. All scientists have to do is patiently continue to follow the trail toward full knowledge. (Unfortunately, unless we become far more responsible in looking after our planet, immediately, we may never complete this journey. However, we will be optimistic and play with this idea under the assumption that we will not self-destruct.)

It makes little difference whether final understanding arrives in the twenty-first or the twenty-fifth century. For the sake of playing with the idea, let us accept this prediction as correct. Our hypothesis is that the human species will someday, relatively soon, reach a level of essentially complete understanding of the basic laws of the universe, and hence comprehension of how the laws operated to produce the physical universe in its present form. That does not mean that we will understand every small detail, only the broad issues, including a great deal about the nature of biological phenomena. The finest details may elude us forever, given that individual decisions appear to play a central role in defining the nature of the details themselves. For example, whether the ozone layer is destroyed may ultimately depend on decisions made by the chemical industry, and those, in turn, rest upon all of us and our consumer habits. Ultimately, the outcome of our decisions is not predictable, according to chaos theory, which recognizes that from tiny causes vast effects may grow in a quite unpredictable manner. But even chaos theory does not deny that we may come to appreciate and understand the broad details of the laws that determine how matter behaves in our universe.

It seems highly likely that humankind, four hundred years after setting aside superstition and a mere two hundred years from the time that experimentation began in earnest, confronts the possibility that many of the basic questions that have for so long been asked of nature have already been answered. Humankind may be standing on the brink of knowing all there is to know about the *broad aspects* of nature, in particular about the forces that act upon physical, chemical, and, very soon, biological systems, at least to levels of detail that make a difference in everyday life. I want to stress that it makes no difference to our discussion whether we consider the state of final understanding to be very near or a hundred or a thousand years in the future. The point is to recognize that progress in the quest for knowledge has been extraordinary, especially during the twentieth century. Now it may be time to ask, What next? What will curious humans do when the answers to their fundamental questions have been found? (If you do not believe we will ever approach a time of full understanding of the fundamental issues, why do we bother to continue in our quest for knowledge?) The human mind has made enormous strides toward understanding the nature of the physical universe, and in time the biological realms of existence will reach a similar level. Isn't that what we have asked for, as a scientifically oriented society that wishes to understand the nature of existence? But once we have found it, what is then left to be done?

Answers may range from solving social and psychological ills to bettering the lot of all human beings, and so on. And that is the point. The curious human mind, in its persistent questioning, appears to be approaching the boundaries of the physical universe on all scales and in all directions of thought. Once we understand the nature of matter and of the forces that act upon it, we will no longer ask traditional questions. The most fundamental ones, such as, What is light? What is heat? What is gravity? What is magnetism? will all have been answered. Richard Dawkins made this point in his treatise on evolution.[3] He points out that the mystery of human existence *is* solved. Similarly, the mystery of magnetism *is* solved. In fact, given the depth of understanding of the four basic forces of the universe, the broad picture lies spread out before us. Details may still be explored, but that is not what concerns us.

How can we say that the mystery of magnetism is solved? Well, we have seen that magnetism is produced by electricity in motion and that magnetism is one aspect of electromagnetism, which, in turn, is one aspect of the electroweak force. There is every indication that this is an aspect of another force described by Grand Unified Theory. Our search for answers to basic questions in diverse fields of human experience, and specifically of physics, has led to an increasingly intimate view of the underlying *unity* of nature, just as Kant suspected it would.

This brings us again to questions related to the origin and evolution of life. Despite continual brushes with religious fundamentalists, whose beliefs are deeply threatened by the notion of evolution, biologists have plunged ahead and built upon what they have discovered. Today we are at a point where microbiologists clone genetic material and have begun a program to interpret fully the genetic code (the Genome Project). This is a major step toward full understanding of the gene, and that, in turn, is a precursor to being able to find practical uses for such knowledge. In the biological sciences we may stand at an awesome threshold not unlike that which confronted students of electricity and magnetism in 1800.

Once upon a time, when Coulomb, Oersted, Ampère, and Faraday made great strides in understanding the nature of magnetism and electricity, it became possible to formulate workable theories for the phenomena. These theories laid the foundation for the commercial exploitation of their discoveries. Commercialization proceeded as soon as the experiments suggested how to use the new knowledge, even if the theories were still in need of revision. Invention of practical devices did not require that the theory be fully understood. It was not necessary for Thomas Edison to understand quantum electrodynamics to build a light bulb. It is not necessary for your utility company to know about the *standard model* relating to physical forces to deliver electricity to your home. Similarly, it was not necessary for Marconi to wait until the electroweak force was discovered before he sent radio waves across the Atlantic.

The same will be true in microbiology. No one needs a perfect understanding of DNA, or all the ins and outs of evolution, to synthesize insulin or find ways to fight a viral or genetic disease. But when essentially complete and full understanding dawns, as inevitably it must, provided we are still around as a species, humankind will not only know how the physical universe operates but will know how biological systems function. When that time arrives (and again it does not matter whether that time is ten, a hundred, or a thousand years in the future), our species will be faced with unparalleled opportunity as well as potential danger. When we understand biological systems as intimately as we now understand, say, the unification of the electromagnetic and weak forces, it may become possible for our species to create anything it wishes. At that future date the manipulation of life will surely become just as pliable to our control as the phenomena of magnetism and electricity are today. The human species will then be able to alter life to suit its fancy: cure disease, create disease, arrest aging, and determine the nature of offspring to the last detail. Such abilities may soon be ours. But then what?

A potential tragedy lurking in the wings may be the lamentable ill-preparedness of our species to deal with the awesome responsibilities it will confront when that day arrives. We may be unable to handle such knowledge intelligently. If we prefer not to think about these options, we can pretend that this scenario is so far in the future that it doesn't matter. On the other hand, the lesson we have learned from physics and astronomy is that once understanding begins to dawn, progress occurs with astonishing rapidity. As we learn more we also appear to grow in our capacity to ask better questions, which are then answered even more quickly. Whether society as a whole can deal with this (exponential) progress remains to be seen.

Our species is traveling at an enormous pace toward a profound and far-ranging understanding of its existence in the context of the physical universe in which we find ourselves. Robert Ornstein and Paul Ehrlich have argued that the world is now changing faster than people can become adapted to it.[4] I suspect that this may act as a natural limit to how long our recent and extraordinary progress in understanding nature may continue unabated, until we slide, perhaps temporarily, into a new dark age. But sooner or later the quest is likely to be resumed. Inevitably all the answers will be found. Of course, this speculation assumes that our species as a whole will survive, and we have no indication that that is inevitable.

For several centuries we have asked difficult questions and the answers have been or are being found. Even if we are not completely happy with the details of many of those answers, there is every indication that the human mind, miraculous organ that it is, is quite capable of finding solutions to the most interesting mysteries, answers that can be tested in the vast laboratory of nature, or in small-scale laboratories on earth.

With progress we learn to design better experiments to search for clearer answers. Inevitably, progress grows more rapid. From this perspective we must confront the possibility that our species will, before long, have learned to understand the essential issues involved. After all, the universe is finite and physical phenomena proceed according to only a small number of identifiable laws.

It is only a few centuries since we began to ask questions in a truly meaningful manner, which then allowed us to develop technology to help discover further answers that had a bearing on physical reality. No longer are our explanations based on wishful thinking, superstition, or religious belief. That is what is extraordinary. For thousands of years it was taken for granted that the explanations for the great mysteries of existence involved metaphysical processes. But we were wrong. Slowly, conscious minds began to suspect that there was something peculiar about those ancient answers. Once the Era of Superstition in a given field of inquiry was left behind it became possible not only to learn the truth about nature but also to learn about the nature of truth. (Perhaps the point is that the age of magic is transcended when everyone learns to do the tricks.)

Explanations offered in terms of passive superstition were invented at a time when the human mind was just setting out on its path toward self-awareness. In those long-gone ages of darkness, human consciousness could barely comprehend its own existence. Like the baby that begins to discover itself, our species once looked around and asked its first questions. When that occurred we will never know; all we can be sure of is that it must have been at least ten thousand and perhaps as many as a million years ago. The first answers our youthful species devised could only be ones that primitive human beings invented and understood at their level of mental development. Their answers provided solace in the face of the vast unknown of existence. But today we live in an era of significant understanding, although it is not obvious that we find much solace in the understanding we have wrested from nature. The universe, after all, was not designed to keep us happy.

When exploration in the realm of physics began in earnest, complete understanding was not far in the future, at least not as measured on the time-scale of human civilization. Will it be any less rapid for the biological and, inevitably, the social and psychological sciences? Will the rapid exploitation of discovery in the practical realm in those disciplines be any less dramatic than was the invention of the electric generator and the dynamo? For example, the Genome Project, when complete, will allow the microbiologist direct access into every aspect of the human gene. Once the DNA has been mapped there may be no holding back. Why even consider holding back, except to avoid the dangers inherent if we consider that our species may be too young to play with these things?

Like a child that plays with matches, we face unknown dangers as we discover the answers to our questions. Curiosity has taken us this far and continues to drive us on, sometimes far too rapidly for comfort. Is there any reason to think that our collective curiosity will fade? On the other hand, how much further can we go? Can we learn to live comfortably, all of us, with the knowledge that promises to overwhelm us? Or will vast segments of the population remain ignorant about what is now being discovered in laboratories around the world?

Wise people of all ages have suspected that something like this might happen, that hell on earth would befall those who ate of the apple of knowledge. The relative mental peace that may once have accompanied ignorance is gone forever. Our species grows ever more uncomfortable with its vast accumulation of data and knowledge. Is this what it means to leave the Garden of Eden, where once we dreamed and dwelt in mythological bliss?

When we finally understand the nature of the physical universe—and that understanding appears to loom very close—we will appreciate the themes and variations that allow for the existence of life as we know it. Then all ancient superstitions will be relegated to the realm of historical curiosities. At that time the species will have to confront a new and far more sobering question: What do we do next?

Already we are forced to ask this as we face the disastrous consequences of rapid technological progress that are embodied in the unrestricted exploitation of the planet's resources. Blindly and unconsciously we destroy the ozone layer, threaten a greenhouse catastrophe, pollute the oceans and rivers of our planet, and overwhelm the ecosystems with billions of additional human beings. Our species already must confront how it wishes to survive in the future. But is it capable of doing so without making conscious decisions regarding the future?

So far human curiosity has been posing questions of nature. Much to our surprise we have found answers at levels no one could have dreamed about a century ago. We have asked questions whose answers lay deeply hidden within natural phenomena, behind the manifestations of nature in all her colorful garb. We have asked why lodestone was magnetic and relentlessly insisted on asking the question again and again. Answers were forthcoming. We were rewarded with knowledge that led closer and closer to the truth that exists beyond appearance. Biologists now ask about the nature of life and take their search into the microscopic realm. They have found DNA and genes and ask how those function. They, too, are wresting the most elusive answers from the matter of which all of us are made.

Now project yourself into the future, in imagination. Again, it matters not whether you think we will have essentially total knowledge in ten, a hundred or a thousand years time. Transport yourself to an era when physicists have seen the deepest truths about the nature of the physical universe, and biologists have fully understood the workings of the genes. Imagine a time when astronomers have uncovered essentially all the phenomena that will ever present themselves to our view.[5] Surely there will always be details to be filled in, but let us look past that time into an era when the details no longer make any difference to the underlying truths that have been revealed. Imagine also that we have not destroyed our species or our environment. What then? What questions will then remain to be asked?

At that time, which may not be as far off as we imagine, we can no longer ask what or why or when. What happens next will be determined by what we *decide*, collectively, to do with our knowledge. The question confronting our species will then become: What do we wish to create?

At the time these sentences were written, Iraq was found to have been constructing a nuclear weapon. The cynic may claim that that is an example of where our scientific knowledge leads. But I am trying to project past an age where regional powers struggle to control neighbors, where our collective consciousness has evolved to a level where human beings recognize that they are one species.

Armed with full knowledge of the nature of space, time, matter, and the workings of the forces of nature, as well as a full understanding of the nature of the gene and of life on earth, what might we wish to do with such knowledge? After all, this level of knowledge is the goal of scientific endeavor. If it isn't, what are scientists aiming at? The point is that the genie of understanding is already halfway out of the magic lamp and is reminding us that we called. We asked and answers have been found. We asked to see more clearly and we learned to look. We wanted to know and now we do (give or take a few details and another decade, or century or so). It took four hundred years from the time we began to ask questions in a serious way and look how far we have come! We rubbed the lamp of knowledge and have actually learned something. Our species has begun to understand nature's deepest secrets.

"What next?" the genie insists on knowing.

What do we want to do with all our knowledge? Consider that as soon as physicists understood $E = mc^2$ someone realized that the insight could be used to build a nuclear weapon. As soon as someone discovered the principle of fusion, physicists tried to figure out how to control it so as to provide endless power. (Success seems elusive here, and why not? In order to create fusion power nature uses entire stars. Can humans do any better?) The key point is that our curiosity has led to answers. So what will our species want to do next, once its curiosity is rewarded? Will we become a planet of nitpickers exploring the privacy of terribly tiny details?[6] Or will we ask new types of questions: those of the form, What do we want to create with our knowledge? or What sort of world do we want to live in?

I believe that when this era dawns, science will become art.

It will be used to create. Knowledge gained through scientific exploration may then be applied rationally to create a new world, much as paint is used to create a work of art on the surface of a canvas.

This image of the future assumes that we will not be in a constant panic to deal with immediate crises of our own making, such as a greenhouse effect or destruction of the ozone layer, or how to avoid a devastating asteroid impact.[7] Instead, our relatively conscious species (*Homo sapiens* or whatever may replace it) will reach a point in its evolution where it no longer explores the unknowns of the physical universe. Humankind will reach a stage where it no longer seeks knowledge about the physical universe because the mysteries that currently surround many of its manifestations, including life, will have been pried loose. Then our species will confront how it wishes to move ahead, in full awareness of its collective powers and its shared storehouse of knowledge. Already there are hints that this is occurring, but usually the forces that drive us ahead are based in crises: of the environment, of population, or some regional catastrophe.

Sooner or later, barring an extinction catastrophe, our species will accept responsibility for its continued existence, a responsibility that today is still pressed into the bosom of a variety of gods, invented at a time when our first questioning about the nature of the universe began, when we created imagined answers because we had not yet learned to approach nature directly. If such a day dawns, our species will make decisions of a creative nature that will not only perpetuate its existence but will raise it to whatever new heights it wishes to attain. Perhaps only then will we learn to venture to the stars.

In our scenario, archaic edifices of belief will crumble into dust because they will be unable to resist the advance of knowledge. They will fade into obscurity. It may take hundreds if not thousands of years for this to happen, but surely it is inevitable. I think that personal rationalizations for existence in terms of expressing a god's will, for example, will inexorably die away. But old habits do not die easily.

At all points along our continuing journey toward that era of understanding we will have the option of deciding how to use our knowledge. What will we do with the wealth of insight and wisdom gathered when our primitive questions have all been answered?

This may be where our species is headed, provided it survives its present headlong rush into a more technological civilization with its exponentially growing accumulation of knowledge about the universe. But even a cursory glance at the scientific literature reveals that what is growing so fast concerns mostly the details. The broad, underlying pictures appear to be in place. These are only softly touched by "progress."

A painter must first learn the techniques of his or her art. Then comes a time to create. Once the musician learns to master her or his instrument, there comes a time to play or to compose. After the scientist has learned the basic skills of the trade, there comes a time to ask questions, to formulate models, and to propose new experiments. But once the answers have been found, then what? Can we expect to do anything but *create* in the context of the new knowledge?

I think humankind is headed for a confrontation with its potential creativity. When that occurs, science will evolve into art, to be used for what we wish to create—together. Human beings already do this through the technological exploitation of scientific discovery, albeit largely unconsciously. But I wonder what will happen when the scientific questions have been answered. (Or do you believe that the National Science Foundation and NASA will go on funding scientific research for thousands or even millions of years to come? If so, does that mean we will never find answers to our basic questions? And if that is so, how do we justify searching for that which will never be found?)

Implicit in this scenario, then, is the awareness that our species will come to understand the nature of physical universe and be able to manipulate it within the bounds of the laws of nature. We will learn to understand the nature of biological existence to the point where we will literally be able to take over from nature. Whereas the physical universe runs according to well-known laws, the biological world appears malleable. The phenomenon of evolution has proven this. Life can take on many forms, which are adapted to specific niches in the environment. But once humankind fully understands the ultimate details, will there be anything to stop us from designing new environments and life-forms to fit those environments? Will there be anything to stop us from turning ourselves into anything we want to become, given only that some natural or self-induced disaster doesn't bring all this opportunity to a premature end?

In the not-too-distant future the human species will have to ask what it wants to do next. Someday we will graduate from that school we currently attend, the one in which we study nature. After graduation we will have to create larger roles to play. Whether we succeed or fail in that upcoming adventure is up to each of us. But is this likely to happen soon? It seems unlikely, because the planet is far from united in regards to how to approach nature's truths. The era being imagined here can only dawn when we have evolved completely out of the age of superstition, a time that may still be far in the future. The trouble is that individually, and certainly collectively, it is difficult to let go of cherished beliefs. However, as many people now warn, potential ecological catastrophe may force us to face this issue if we wish to survive. It remains to be seen whether we will be able to transcend the beliefs that bind (and that separate) cultures.

Once upon a time lodestone was a complete mystery. Then Peter Peregrinus struggled to put in writing what he knew to be true about its properties. Three centuries later Gilbert sorted out superstition from what was known about magnetism. Within the next few hundred years, experimental physics came into being, allowing human minds to approach and then to uncover the truth about magnetism.

I have used the subject of magnetism to illustrate a broader theme: the evolution of science from superstition to certainty. Our story of the quest to understand magnetism is but one of many similar examples that reflect how the human mind progresses in its search for answers to the mysteries of existence. Relatively soon, what were once burning questions were mysterious no longer. Today, at the beginning of the third millennium, we stare at the possibility that a unified theory may be found that can account for all aspects of the simple complexity of the universe. Given a beginning to our universe, defined by a small number of laws of nature, it is clear that enormous complexity can be generated. After all, we are here, now.

The most awesome fact may be that the human brain has emerged from the conjuring of evolution to the point where it has learned to comprehend its own existence, to fit together so much of the jigsaw puzzle of its world. Despite cries of caution that we cannot fully know our origins, it appears that we are well along the path toward a profound, and perhaps even complete, description of our universe and life on earth. If this is so, before long we will cease to ask traditional questions. They will have been answered.

Our species may be confronting the dawning of an era of creativity, an era during which the human species will have to decide what it wishes to do for eternity, right here on earth, not in some imaginary world hereafter, not in a space colony, not on Mars, but right here. Our species, as we learn the most intimate secrets of the universe and everything in it, will surely begin to create. In a sense, our grandchildren's grandchildren—who knows how many generations removed—may yet become as gods on this planet, with the power to determine what to do next and how to get there. But where will that future lie? Has our species evolved far enough in the psychological and social realm to make use in creative ways of the enormous fund of knowledge it has already gleaned from its explorations of the physical and biological realms? Or are we precariously poised with one foot in the era of superstition and the other in an era of understanding?

Perhaps it is mere entertainment to give any of these issues more than a passing thought. But whatever one's point of view, we appear to have come further along the path toward understanding than we may be comfortable with. In taking a long look at that path, we have brought to light the story of magnetism to illustrate how progress was made in at least one area of inquiry. Similar things are happening even now in every realm of scientific curiosity. After all, scientific inquiry has been demonstrated to be a wondrously powerful way to arrive at the truth about the nature of our universe.

NOTES

1. Stephen Hawking, *A Brief History of Time.* (New York: Bantam Books, 1988), p. 156.

2. See Hawking, Ibid., for an optimistic appraisal of how close we are to a full understanding.

3. Richard Dawkins, *The Blind Watchmaker*. (New York: W. W. Norton, 1986), p. ix.

4. Robert Ornstein and Paul Ehrlich, *New York, New Mind*. (New York: Doubleday, 1989).

5. This future has also been described by Martin Harwit in his book *Cosmic Discovery*. (Cambridge: MIT Press, 1984).

6. Ibid.

7. See, for example, Gerrit L. Verschuur, *Cosmic Catastrophes*. (Reading: Addison-Wesley, 1978) and "The End of Civilization?" *Astronomy* (September 1991): 50. Also Clark R. Chapman and David Morrison, *Cosmic Catastrophes*. (New York: Plenum Press, 1989).

n 15 n
The Wages of Curiosity

All animals feel Wonder, and many exhibit Curiosity. They
sometimes suffer from this latter quality.

Charles Darwin, *The Descent of Man*

C HILDHOOD memories of
my fear of lightning came
flooding back to me as I considered the question "What is magnetism?" It began to look almost self-evident that, like so many
other questions in the history of the natural sciences, explanations for magnetism were originally rooted in superstition. Yet,
over the millennia, curious people managed to transcend superstition as they began to find answers that had a bearing on
reality; that is, the way the universe really is, as opposed to the
way someone might have wished it to be. How they did so in
regard to magnetism is the theme of this book.

Significant and rather sudden progress in understanding this
phenomenon during the early nineteenth century marked a great
watershed in human thought, a divide that signaled the end of
a primitive fascination with, and reliance upon, untested beliefs
for providing answers to fundamental questions. That was the
time when experimentation began in earnest. The breakthroughs in understanding magnetism that resulted were accompanied by giant steps forward in astronomy and in physics
in general, and set the scene for our modern scientific era in
which experiment and observation (which go hand-in-hand) are
recognized as the primary means for getting at truth. Only then
did it become possible for the rational mind to move toward a

meaningful confrontation with the nature of the physical universe. The rational mind had at last been forced to face the way things are rather than to dwell on how a variety of dogmas insisted they should be.

There is a pattern of progress that underlies our efforts to understand mysterious phenomena, a pattern that can be recognized by looking at historical examples, and at human action today.

The beginning of a search for answers is almost invariably rooted in myth and superstition. Then someone challenges the myths and, if possible, begins to experiment to obtain data about the phenomenon in question. Only then does the door to understanding begin to swing open.

Scientific research is based on finding testable and meaningful answers to questions about the nature of physical phenomena, and the search for truth passes through distinct stages (see Appendix for an outline of this process). These are remarkably akin to those manifested in ancient societies, and by individuals in their dealings with day-to-day experience, especially in the case of children. Each question asked is first dealt with in terms of fictional constructs, beliefs, superstitions, or myths that strike a chord within the human psyche. But that does not make the explanations "correct," at least not if you want to know the truth. History has made it clear that human minds were never really satisfied with such answers. That is why we continue to question and, very gradually, to find answers that pertain to *reality*. In the process something occurs that most of us take for granted. We begin to learn facts about the nature of reality. These can then be exploited through invention and technology, which, in turn, influence every aspect of our existence on earth.

The days of dependence on superhuman entities to account for natural phenomena are essentially behind us, despite the fact that many people still cling to such explanations when it comes to accounting for life. My thesis is that the transition from superstition to modern science as a way of reaching truth was marked by the ferment of intellectual activity in the first half of the seventeenth century. It was then that magnetism, electricity, gravity, heat, and light all became subjects for experiment. It was then that the groundwork was laid for the study of the planets and stars, earthquakes, volcanos, and the nature of living things. To satisfy curiosity, human beings began to rely on observation and experiment rather than beliefs.

The story of magnetism illustrates the expression of this syndrome. The pattern of progress can be recognized in the study of astronomy, biology, meteorology, geology, electricity,[1] and other areas of curiosity that were once haunted by supernatural entities. This is particularly true of astronomy, where a gaggle of gods was once believed to control the motions of the stars, sun, and planets. It was true in geology, where the rumblings and upheavals of the earth were the work of powerful spirits. It was so in biology, where such spirits breathed life into inanimate matter to bring us into existence. And in electrical phenomena what god was more powerful than Jupiter, tossing about lashings of lightning bolts?

We can now look back from our safe perspective in time to recognize that the invocation of gods reflected only the most primitive attempts to explain nature's phenomena. This need to create a model appears almost instinctive. The creation of an imaginary analog—a model—of some aspect of reality represents the mind's need to believe it is in control. I am unaware that societies or groups of individuals who resort to beliefs to guide them would invoke images of chaos and confusion, which is what the universe, to a major extent, is really like. Tornados do not respond to prayer. Hurricanes are not swayed by offers of sacrifice. Stars explode whether or not surrounding planets are inhabited by sentient beings. Entire galaxies are torn asunder by cosmic violence no matter how many inhabited planets may dwell in their realm. Earthquakes wreak their damage ignorant of a quota of miracles that allow certain individuals to survive a near disaster while a neighbor is crushed by falling debris. The universe is a frightfully vast and violent place, and individual human beings an infinitesimal speck in the cosmos.

To succeed as an explanation, a model for some aspect of reality must appear reasonable and comprehensible. That is why the earliest models for natural phenomena involved superstitions or myths that were derived from the simplistic concepts that could be appreciated by the generations who formulated the explanations. This relates to the Jungian concept of the existence of a collective unconscious, to which our myths resonate. That is why today's models for reality, our modern explanations for phenomena such as lightning or magnetism, would be incomprehensible to the ancients. Scientific discoveries about the nature of physical reality find no resonance with our psychic depths. Why should they? Instead, what we have found at its deepest roots defies intuition. Yet we appear to be able to appreciate the nature of our universe, even if it is unlike anything we might have expected to find. It certainly is unlike anything that any ancient belief system could have foretold.

And so the quest goes on. Our fund of concepts (and language itself) has evolved so we can deal with the nature of the phenomena that lie at the root of the truth we seek. Above all, the human mind learned to make models of reality, models that could be tested against that reality. Sometimes a model may involve a real, physical simulation of a phenomenon. Thus a powerful electrical generator can be used to produce controlled lightning. It can be used to create a model lightning flash. And what better example of the value of a model, of humankind's approach to the truth, than the modern picture of the atom. Forming a reasonably correct idea of the atom and of the physics of its behavior allowed a hydrogen bomb, a small-scale version of a star, to be built. The fact that the bomb worked meant that the scientists understood one of nature's great secrets well enough, had formed an accurate enough model of the structure of matter, to make the bomb explode. The measure of a success of a model rests in what we can do with it. Does it allow us to mimic nature, or is it merely a figment of our imagination?

The goal of science is to be able to understand aspects of reality so well that scientists can make models, small-scale versions of reality, that closely approximate the original. In this sense, a model is a picture or simulation of a phenomenon, an analogy constructed on a scale we can deal with to demonstrate insight. Models may be of a theoretical nature, consisting of equations and scribblings on paper. They may be found inside computers, where electrical pathways controlled by the program allow equations to be solved to reveal answers. In all cases, models are meant to explain observed phenomena and, above all, to predict new ones. The better they do so the better the model. Today's models in science vastly supersede any that have gone before. By any measure of success, scientists have become extraordinarily good at approximating nature's phenomena.

Science appears to have reached a point where further discovery tends to refine, not to overturn, what we know to be true. This statement rests on the fact that we have learned to simulate nature's phenomena to a remarkable degree of accuracy. For example, nothing that will ever be discovered in physics will undo the fact that an atomic bomb wiped out Hiroshima. The model of the atom that allowed the bomb to be built was an excellent approximation of the truth. Further progress in understanding the nature of matter may offer refinements to the theory, but the model of the atom that allowed a bomb to be built was close enough to the truth to be regarded as an accurate model; otherwise the detonation would not have occurred.

It can be argued that from now on much of science may concern itself with the details involved in getting even closer to the truth. This suggests that in the not-too-distant future the desire to get closer to what is already at our fingertips may be of interest only to the experts, if we are not already there in some of the physical sciences. (A version of this point of view has been argued by Harwit in regard to astronomy.[2])

These comments apply to sciences other than physics. For example, biologists are very close to understanding the broad basis of life. This can be seen in the way they manipulate its building blocks—genes and DNA molecules. Fiddling with the stuff of life is only possible if one already has a very good model, one that is close to the way things really are.

How close we are to the truth can only be measured by the extent to which we can simulate natural phenomena under controlled conditions, whether in the laboratory or in theoretical calculations. In the past, a new concept might have altered the way we understood magnetism or electricity. And thus progress was made. But not until the late nineteenth and twentieth centuries did the models become such good approximations of reality that we gained the power to use the knowledge gained from experiments to exploit the secrets of nature. That allowed the invention of electric generators and motors, for example.

The message of scientific progress in the late twentieth century is that there is very little that scientists cannot already do, or at least conceive of doing in the relatively near future. Their models, their approximations to reality, are working. Curiosity has inexorably led the explorers to approach very closely the nature of truth in many areas of inquiry; in astronomy, physics, chemistry, biology, and geology. It is what we plan to do with this knowledge in the next millennium that should concern us, as was suggested in Chapter 14.

From this perspective, then, our saga illustrates the process the human mind went through as it sought answers to a profound question: "What is magnetism?" It is significant that rapid progress toward finding an answer was in no small part due to the fact that organized religion never opposed progress. The study of magnetism (and its twin, electricity) was never fettered by religious dogma, as was the case for the study of the origin of human beings, for instance. This meant that progress was always and only dependent on human curiosity and ingenuity, especially once the era of experimentation began. The same cannot be said of other areas of science, such as the seventeenth century investigation of the motion of the planets, where Galileo's confrontation with the Church became a symbol of that era of change; or the nineteenth century discovery of evolution, in regard to which ill-informed fundamentalist reaction persists until today. None of this is surprising, of course, since no one likes to let go of cherished beliefs, especially if they form a cornerstone for one's mental peace. But then the search for truth and its final comprehension never did guarantee peace of mind after the truth was found.

Progress toward the understanding of natural phenomena is possible only when two conditions are met: curiosity must be allowed to seek satisfaction unfettered by authority, and inquiry into the truth underlying appearances must be aided by the means to perform measurement. These are essential for understanding nature and allow scientists to approach the quantitative aspects of reality. Measurement provides the data the mind can apprehend as it seeks to identify patterns that reveal the truth underlying appearance. That is what the scientific endeavor, and perhaps all of human inquiry, is about: to determine what lies beyond appearance, beyond the manifestations of everyday existence.

Success in uncovering that truth, and understanding its meaning, is a measure of progress, although progress is only meaningful if it brings us closer to understanding the way things *are*. Not all experiments in the quest for answers lead to immediate success. We hope that successive experiments bring us closer to the truth. Whether we are getting closer can only be measured by how well we can simulate reality. In science the judge of this progress is experience. We know we are making progress when the pieces of the jigsaw of existence begin to fit together meaningfully enough to allow us to recognize the picture, and use what has been learned to move ever closer to the underlying truth we seek. We know we are close to the truth when we use what the picture reveals to build our own version of the image (the model) and find that it can be made very much like the original.

Herein lies the difference between the ancient reliance on superstition and the use of scientific experiment as a way for getting at the truth. You could do nothing with superstition: invent nothing, construct nothing, and predict nothing. That remains true today, especially in astrology, a collection of superstitions to which people cling no matter what the evidence of their experience. The goal of science, however, is to predict correctly, where correctness can be tested only against reality.

Meaningful scientific knowledge can be used to create new things, new ideas, new theories, and better models. These may then lead us even closer to the truth. It may be moot as to whether we will ever find absolute answers to our questions. What is important is that we get close enough so that for all practical purposes we can use our knowledge to simulate reality to a degree where the differences are barely noticeable, except to the most fastidious eye. For example, if you want to understand why a star shines it may not be sufficient for you to be able to explode a hydrogen bomb. You will want to build your own star, even if it is in the heart of a computer as a set of instructions that bring imaginary masses of gas close together so you can watch it evolve in numerical form.

In seeking to answer my fearful questions about thunder

and lightning, my parents instinctively resorted to an archetypal behavior pattern that had its roots in ancient times. In response to our need to know, in order to feel more secure about the world around us, we were first given invented answers. As long as answers provided some verisimilitude of relevance, early explanations had to be couched in terms that lay within the primitive mind's limited experience to understand. In my experience, searchlights were quite capable of producing sudden beams of light and empty oil drums resonated sonorously when struck. My later experience revealed the shortcomings of this model. Perhaps that, in the end, is a measure of whether we accept a myth, superstition, or theory to account for certain phenomena. If a myth satisfies our need, we will accept it. But there are always curious souls who challenge the dogma and probe beyond. Therein lies the birth of the scientific quest, a quest whose outcome is not guaranteed to cause us to sleep easier. After all, natural phenomena play themselves out with no heed for our personal existence. Lightning is not produced by a man operating a searchlight. Lightning is a lethal bolt of electrical discharge between the clouds and the ground, no matter who shelters under the tree that may be struck.

For millennia the phenomenon of magnetism appeared magical. As long as magic was what society wanted, there could be no progress in understanding this remarkable phenomenon. As it turned out, understanding magnetism and being able to create it at will proved to be fundamental to setting the scene for the creation of our modern world. Without knowledge of, and hence control over, magnetism we would have no electrical generators, no motors, no power for industries or our homes, no telephones, radio, TV, radar, or computers. We would also know very little about the nature of our universe, both on the atomic and cosmic scale, because it is with a multitude of electromagnetic waves such as gamma rays, X-rays, ultraviolet, light, infrared, and radio waves that we probe into the universe to discover where it is we reside and what the world is made of. These waves would not have been discovered, and certainly not understood to the degree of being able to exploit them, without understanding the nature of magnetism.

NOTES

1. Although not specifically noted, the early phases of progress in the study of electricity were reported by Alfred Still in *Soul of Amber*. (New York: Murray Hill Books, 1944).

2. Martin Harwit, *Cosmic Discovery*. (Cambridge: MIT Press, 1984).

APPENDIX
The Pattern of Progress

Does a field make progress because it is a science, or is it a science because it makes progress?
Thomas Kuhn, *The Structure of Scientific Revolutions*

USING the story of magnetism as an example, I have argued that the first mental constructs invented by human beings to explain natural phenomena fell into the category of superstition. This type of thinking held sway until someone confronted a specific issue head-on in the sense of seeking an answer tied to reality, an explanation that had meaning as defined by the nature of the universe external to our minds, as opposed to an explanation created by our hopes and expectations. This phenomenon is itself natural, because at the dawn of civilization answers to difficult questions could only be *invented*. Methods for approaching nature through direct experiment were inconceivable to the primitive mind. Why otherwise did it take until the seventeenth century before the scientific method was discovered? Thus it was that astronomical phenomena such as the daily motion of the sun, moon, and stars, the phases of the moon, eclipses, and shooting stars were at first explained in terms of fiction, and then only after the brain had evolved to the point where it could create fiction, no small accomplishment in itself. Once that point was reached, the newly emergent mind found solace in imagining entities such as a sun god in a chariot racing across the sky, a sun god who was actually just as mysterious, although more personal, than the phenomenon it was meant to explain.

Our story has also suggested that at least in the realm of research into the nature of magnetism the search for answers to a basic question passes through distinct phases. Together with the speculations concerning the future in the previous chapter, it is possible to recognize a pattern of progress in science. Broadly speaking, several phases and stages can be defined as follows:

Phase I: From superstition to uncertainty

Stage 1. The era of superstition
Stage 2. Clearing the decks
Stage 3. Uncertainty

Phase II: Grasping at reality

Stage 4. "Primitive" experimentation
Stage 5. Working models developed
Stage 6. The invention of measurement techniques

Phase III: From discovery to the primary paradigm

Stage 7. Initial discoveries
Stage 8. Emergence of a primary paradigm
Stage 9. Elucidation of theories

Phase IV: Parting of the ways

Stage 10 a. Technological exploitation
Stage 10 b. Continued research

Phase V: Approaching the truth

Stage 11. Paradigm shift
Stage 12. Synthesis
Stage 13. The theory of everything?

Phase VI: The age of creativity

This scheme outlines the way progress was made in the study of magnetism and in most of those realms of experience that are the subject of other branches of science, especially those that had their origins in ancient history. By offering it for your

consideration I am acutely aware that many of my colleagues
will dismiss the scheme as simplistic; yet this pattern seems clear
in the study of subjects such as magnetism and the larger realm
of astronomy as well. Certain modern sciences, however, such
as radio astronomy, with their own technology and realm of
inquiry, can be seen as having come into being full grown, as it
were, in Phase III, for example. You may try to place your fa-
vorite discipline in this scheme, just for fun. For example, where
are biology, chemistry, psychology, economics, or the various
social sciences? How are they likely to move ahead in the
future?

Various branches of science have always been, and continue
to be, at different stages in the evolutionary sequence proposed
here. In the case of the study of planetary motion one may
argue that an era of confusion and uncertainty lasted nearly
two thousand years, as astronomers tried to make epicycles work
as a means for explaining how the planets orbited the earth.
Until Copernicus, Kepler, and Galileo between them cleared the
decks of confusion, astronomers were unable to progress toward
understanding planetary motion. Even then, years of uncertainty about the nature of astronomical phenomena lay ahead,
an era that ended in the midnineteenth century only when an
instrument (the spectrograph) was invented that allowed the
detailed nature of light to be studied.

Within a given science not everyone may be at the same
stage. In the study of magnetism during the midnineteenth
century, the British and European schools were often far apart
and hence at different points on the spectrum of progress.

I have divided the scheme into phases and stages because
within a given phase there are distinct stages that we must pass
through before a major change in understanding can take place.
Let us look at these in a little more detail.

Phase I: From Superstition to Uncertainty

This phase may be recognized across almost all areas of human
curiosity. It is a phase in which many subjects of inquiry are
still embroiled today. During this phase we inevitably move away
from the comforts of superstition to an era of uncertainty dur-

ing which no answers are available. We could argue that the "science" of economics is an excellent example of one field of inquiry that still dwells in this phase, at least if we are to recognize the significance of the label used by a recent United States presidential candidate referring to another's fiscal plan for the country. He called it "voodoo" economics, referring to its basis in superstition. Whether or not it was meant to score political points, the voodoo label touched at the root of our efforts to answer complex questions, especially before we are ready to do so.

Stage 1. The Era of Superstition

During this era the emerging human mind seeks succor in superstition in its quest for answers. Initial explanations of natural phenomena usually reflect fear of the unknown. To the primitive mind, many, if not all, natural phenomena appeared magical. The existence of lodestone, a naturally occurring magnetic form of iron ore, fell into this category. The lodestone was magical because it attracted other pieces of iron over a distance and did not require the intervention of any physical agent to establish a connection. Consider as another example the fact that in ancient times the motions of heavenly bodies were believed to be caused by the action of many different gods. The Egyptian god Ra (the sun) was believed to sail across the sky in a boat. In many primitive cultures each of the planets was associated with its god, a fantasy that continues to enrapture astrologers. Similarly, both static electricity caused by rubbing amber and the strange property of lodestones, which we now call magnetism, were thought to indicate that souls inhabited these objects. Other natural phenomena, such as lightning and thunder, expressed the action, usually wrath, of specific gods. In many societies such deities are still believed to be in control of natural phenomena, including the origin of life. It is not surprising that many superstitions, in particular those involving one or more gods, were invoked to deal with the terror that filled primitive hearts and minds in confronting the unknown.

Stage 2. Clearing the Decks

The first era is brought to a close when one person or a group of thinkers examines what is unambiguously and demonstrably known about the phenomenon. This first critical examination of beliefs helps sort out the wheat from the chaff. If anything of value is found, it may form the rudimentary data base upon which the pioneer will build. This era may encompass only a single lifetime, because an individual is literally "tidying up" the welter of superstition in which the human mind has become immersed over the past millennia. Subsequently, the collective mind slowly becomes prepared to undertake the journey toward greater understanding.

Stage 3. Uncertainty

An era of uncertainty inevitably follows clearing the decks, because once suggestions or other primitive beliefs have been exposed as not being relevant to the nature of reality there is nothing to replace the old notions. Depending on the nature of the question, this stage of uncertainty may last for centuries. This was the case for magnetism. Despite the fascinating studies precipitated by Gilbert's massive work, especially in Britain and France, two hundred years would pass before meaningful quantitative experimentation could begin. During that time uncertainty about the nature of magnetism reigned supreme, at least in the minds of those who cared.

Phase II: Grasping at Reality

During this phase simple experiments lead to the formulation of working models, models that are inevitably shown to be wrong but that serve the function of laying the groundwork for the subsequent phase, which involves more sophisticated experimentation.

Stage 4. "Primitive" Experimentation

This stage is the first time that ideas can be tested by experiment, although these experiments are of necessity simplistic and primitive if for no other reason than the necessary equipment has yet to be invented that will bring experimentation onto a more certain footing.

Stage 5. Working Models Developed

Relatively simple experiments allow initial models to be built or theories of the phenomenon in question to be formulated. These theories have some bearing on reality, as it were, unlike the superstitions that they supplant. Seen from the perspective of hindsight, they usually appear not only simplistic but also wrong. Nevertheless, this critical phase produces ideas that create feedback that allows further, more sophisticated experiments to be designed.

Stage 6. The Invention of Measurement Techniques

The development of models and new concepts that grow out of the first experiments leads to the invention of a measurement technology without which the study of the phenomenon in question cannot proceed. Sometimes the means for measurement may be invented in an apparently unrelated manner. The key issue is that the means to make quantitative measurements is fundamental if progress toward real understanding of the phenomenon is to be achieved. Herein lies the essence of scientific endeavor. Without the ability to make measurements the path of progress is effectively blocked.

Phase III: From Discovery to the Primary Paradigm

Now the scene is set for real progress. Through clever invention, scientists create a system that allows the quantification of their experiments. Data are gathered. This is essential if theories related to reality are to be formulated. The addition of quantity to describe nature sets the scene for allowing theory to be tested against reality.

In the case of the study of magnetism, this phase was filled with opportunities both taken and missed. Whether a discovery was made depended on whether the mind was prepared to make it. Thus even great scientists totally overlooked certain phenomena presented to them during experimental situations because they were searching for something else. At the same time, others, with more open minds, were blessed with accidental (serendipitous) discoveries, or at least discoveries that we tend to regard as accidental when we look back on them.

Stage 7. Initial Discoveries

This begins in earnest when the instruments that facilitate quantitative experimentation have been invented, and that is usually no small achievement in itself. Imagine for a moment handing a teenager a piece of lodestone and asking her why it attracts a paper clip. Where would she begin her quest, especially if going to the library to look up an answer is out of bounds? She would have to go through the steps that the scientists of centuries past went through, which includes inventing the instrument for measuring the strength of magnetic force.

In this Era of Discovery, rudimentary models are modified into theories that begin to make sense and accord with physical reality. Initial experiments may have been driven by sheer curiosity, or by the hope of finding some previously overlooked aspect of nature, but now experiment allows for the quantification of data, a step that is essential to progress. Such data may allow other researchers to formulate theories or discover laws that will allow the roots of the phenomenon under question to be recognized.

Stage 8. Emergence of the Primary Paradigm

As a result of the collection of data, fundamental aspects of the phenomenon are recognized, and explanations for its existence begin to define a broad picture. This is the Era of the Primary Paradigm, that is, the first broad explanation that serves to describe the phenomenon under consideration. This paradigm then directs the course of further research, and the paradigm may, in turn, be falsified by new data. The possibility for falsification of the paradigm, or any theory that is formulated within its bounds, is intrinsic and essential to scientific endeavor. To propose a theory that is not falsifiable is to revert to superstition and belief.

The phase that involves the emergence of the primary paradigm may endure a century or more and it sets the scene for what Kuhn has described in his thesis, *The Structure of Scientific*

Revolutions[1]: When the weight of evidence becomes sufficiently great, a paradigm shift may occur and redirect our thought. Such shifts can be very disturbing to individuals and to cultures as a whole. To let go of the paradigm is to descend, once again, albeit temporarily, into the realm of uncertainty, unless of course a new paradigm is available for immediate adoption.

Stage 9. Elucidation of Theories

Now theories, structured in accord with the primary paradigm, become formalized in the language of science: mathematics, physics, chemistry, etc. History shows that often new languages, such as calculus or symbolic logic, need to be developed for research to move ahead. Clearly not all areas of science have reached this phase.

Phase IV: Parting of the Ways

As soon as enough is known about a subject, even if the detailed theories are far from complete, there comes a parting of the ways. On the one hand, research scientists continue in their quest for truth, but others, with a more technical or commercial bent, will exploit the discoveries in the form of inventions that can be sold to society. There is thus a branching into a technological route and a pure research direction.

Stage 10a. Technological Exploitation

One path involves the exploitation of practical and theoretical discovery. For example, the development of the electric motor and dynamo was possible as soon as certain critical laboratory experiments had been performed, even though neither electricity nor magnetism had yet been accounted for by widely accepted or even well-understood theories.

Stage 10b. Continued Research

The second path continues in the direction of the truths that underlie natural phenomena. This is the realm of pure research, one that many lay people find hard to justify. "What use is your research?" they will ask. Who can answer that question ahead of time?

These two paths, technology and pure research, are not necessarily well separated, nor does the split occur at a unique moment. The pursuit of knowledge is like a tree sprouting branches, which leads to the development of a host of technologies appropriate to the various phases of development.[2]

Phase V: Approaching the Truth

We now come to the phase in which much of modern physics appears to dwell. Despite the abhorrence some people feel for the idea, it appears that we are well on the way to significant and possibly complete understanding of the nature of the physical universe. After all, its actions have been found to be based upon a finite number of relatively simple laws that are apparently capable of being understood by human minds. Thus despite certain arguments that have been aired in the past regarding the essential impossibility of the human mind to understand its own existence, it appears that at least the physical aspects of that existence can be appreciated and understood.

Stage II. Paradigm Shift

A paradigm shift is a fundamental change in the basic model that the scientist has of the phenomenon in question. A change in one's perception of this world view, as it were, may occur at any point along the path toward greater understanding. The paradigm shift will tend to move us closer to the ultimate nature of reality. The shift is forced on us because nature insists that we learn her lessons. Several shifts may occur, and it is moot whether more than two or three are possible in any given field of study. The point of any paradigm upon which we rest our understanding of nature is that it serves very well and is, by definition, a good approximation of the truth. Thus it is difficult to imagine that many fundamental shifts can occur, because the closer we get to the truth the smaller our excursions into new aspects of theorizing, model making, or formulating of paradigms is likely to be. Just like someone who pictures walking from point A to point B as the continual halving of the remaining distance, the closer we get to the end the smaller our adjustments will have to be. In other words, there is every indication in the progress of many of the physical sciences that full understanding is possible and many other sciences are not far behind in their quest.

Stage 12. Synthesis

In this era new theoretical understanding leads to the establishment of connections with other fields of knowledge, either restricted or general. Underlying the pursuit of physics, for example, is this search for a synthesis of basic forces into a unifying whole. Thus several of the basic forces that we know influence matter in the universe are already being related in testable theories, as was pointed out in Chapter 10.

Stage 13. The Theory of Everything?

Whether a Theory of Everything will be found only time will tell. Apart from the physical sciences, it is not apparent that we have reached this era in other areas of rational inquiry. We may argue that it is a wild if optimistic hope even for these sciences. Yet the lesson learned from following the quest to understand magnetism is that when we ask questions we do find answers, especially answers that are meaningful in being related to the nature of reality. Given the persistence of human curiosity, and given the evidence of the historical record, there is every reason to believe that we may well find the deepest answers that can ever be found if for no other reason than that the nature of the universe is such that those answers are intrinsically rational and even relatively straightforward, even if it took our species many millennia to recognize this to be true.

Phase VI: The Era of Creativity

The possibility of this future scenario was discussed in Chapter 14 and is based not just on the evident progress made in the quest to understand magnetism, but on the recognition that this path toward understanding does show an ordered pattern of advance toward comprehension. Then, once our species, or at least those representatives who spend their time seeking understanding, discover the essential secrets that underlie physical and biological existence there may be a simple choice to make. Do we continue to nitpick at the details, or do we use what has been found in a creative manner for the general benefit and well-being of our species and life in general?

Now let me summarize how the study of magnetism passed through the various phases of progress outlined above.

Phase I. From superstition to uncertainty: (1) The era of superstition regarding the nature of magnetism lasted at least twenty-four hundred years, from around 800 B.C. to A.D. 1600, when (2) Gilbert attempted to clear the decks of wild beliefs. Already Peter Peregrinus, in 1269, had made some attempt to formalize what was known about the first magnet, lodestone. (3) Gilbert's work placed scientists interested in the subject into a state of relative uncertainty because testable models of what caused magnetism had yet to be developed. However, it is likely that not many of them would have appreciated how uncertain they were! Of course the enterprise we now call science was only just coming to life, and experimentation on the nature of magnetism would proceed very slowly for the next two hundred years.

Phase II. Grasping at reality: (4) Otto von Guericke's 1660 invention of a simple generator of electricity allowed some useful experimentation into the nature of electricity to begin, but as yet no controlled form of magnetism was available for research. (5) In the seventeenth and eighteenth centuries a number of working models for electricity, including those involving two fluids, were suggested. (6) It was the invention of the torsion balance by Coulomb in 1785 that set the scene for making quantitative measurements of electrical and magnetic force. Shortly thereafter Volta constructed the first "battery," a voltaic pile, which provided adequate amounts of electricity for experimentation with both electricity and, as it turned out, magnetism.

Phase III. From discovery to the primary paradigm: (7) The era of discovery burst forth in the early nineteenth century with Oersted's recognition of the electromagnetic effect of a current in a wire, Ampère's observation that magnetism was produced by electricity in motion, and Faraday's discovery of electromagnetic induction. (8) The primary paradigm that emerged was that electricity and magnetism were intimately related. (9) This relationship was formalized by Maxwell and others into a system of equations that remains in use to this day.

Phase IV. The parting of the ways: (10a) The exploitation of discovery in the commercial sense began almost as soon as electricity and magnetism were shown to be related, in particular after Faraday demonstrated that electricity could cause mechanical motion, and that motion of a magnet near a wire caused an electric current to flow. The motor and the dynamo grew out of these insights and would form the cornerstones of modern technological development.

(10b) At the same time, continued research by Faraday, Hertz, Zeeman, and many others produced breakthroughs that were exploited for further research into, for example, the nature of magnetism in space, and, in the case of Hertz, into more technological development; viz. radio, television, and radar.

Phase V. Approaching the truth. (11) Ever since Einstein realized that relativistic concepts were of consequence in the observable universe, a great deal of effort was put into the search for underlying unifying principles that would link the four basic forces of the universe. One of these forces is electromagnetism, the product of the wedding of electricity and magnetism with a strong dose of time thrown in for good measure. (12) By the late 1960s physicists made more progress toward synthesis, so that now we find that electromagnetism has, in turn, been wedded to the weak force in the electroweak theory. A paradigm shift occurred that involved a unified field theory. It now appears that only time is required to spell out the details. (13) Looming on the horizon is the Theory of Everything whose nature is hinted at in the studies of fundamental particle physics and cosmology.

Phase VI lies in the future and its details cannot be predicted. It is likely to involve a vast and broad synthesis with other realms of human curiosity. This will surely bring to the fore human creativity as our species makes rational and constructive use of all the knowledge gained through centuries of scientific exploration. It will be a time to ask what we will want to do with our knowledge. How this question is answered remains to be seen.

I could construct a complex table to show how several sub-branches of various sciences must inevitably trace their own paths through these phases of progress. For example, with the discovery of radio waves and the subsequent development of radio technology came the fortuitous discovery by Karl Jansky in 1933 that the Milky Way was a source of radio waves. While this discovery could be said to have occurred in Phase III–Stage 9 (elucidation of theory) or Phase IV (parting of the ways) of the study of electricity and magnetism, the subject of radio astronomy itself was born into its own version of Phase I–Stage 3 (an era of uncertainty), skipping, we will admit, the terrible throes of superstition. During the first decade or so of radio astronomy, its practitioners dwelled in a world of uncertainty because no one had any idea of why astronomical objects emitted radio waves.[3] It required the invention of better and more sophisticated equipment (Phase II–Stage 4), which included the world's largest radio telescopes in the United States, Australia, and Great Britain, before the era of discovery well and truly began in the 1960s. Today much of this discipline is no more or less than a part of the larger subject of astronomy. Any future paradigm shifts are likely to involve the most fundamental questions; for example, those concerning cosmological issues related to the origin and evolution of the universe. Is space really expanding away from a minute point that defined the Big Bang (the current paradigm), or is some other explanation needed to account for the many peculiarities that still stand in the way of an intellectually satisfying synthesis?[4]

When one looks too closely at any one branch of science, or its subdisciplines, the boundaries between the phases of progress will blur and disappear. As an exercise in imagination, you are invited to decide for yourself where other sciences and pseudosciences may be placed in the spectrum of progress. For example, astrology is clearly rooted in an era of superstition and will, by definition of the nature of astrology, forever remain there. But where are sociology, psychology, and economics? Wherever we place them now, there is every hope that with time these disciplines will move forward to approach their own great synthesis.

Every topic of (scientific) inquiry initiated in ancient times necessarily involved models that today seem primitive, much as a child's view of the world may seem rudimentary and naive compared with that held by most adults. It is hardly surprising that our culturally shared views of the nature of the universe should have been simplistic, especially when we realize that it is only over the last five thousand years or so that the human mind began to question the nature of its own existence, or at least started to record symbols and images that suggested that it was struggling with these fundamental questions. Stonehenge, the pyramids, rock paintings, and cuneiform script on ancient tablets all attest to this.

For humans to discover truth, nature had to be approached directly, but the methodology for doing so through experiment did not emerge until 1600, the era of Gilbert, Galileo, and Kepler. Only then did superstitions in many areas of human inquiry begin to be called to account. Inevitably they fell away, even if for several more centuries the weight of public opinion remained unaffected by the scientist's radical new way of interogating nature.

When we look back on the evolution of the scientific approach to nature's secrets we recognize that two conditions had to be fulfilled for progress to be made. First, individuals had to be personally interested in the outcome of their quest, and second, their search had to be carried out unhindered by doctrine or authority. The corollary is that whenever a new approach to finding answers threatened the previously held beliefs of large segments of the population, as represented by its religious edifices, for example, progress was impeded.

Today, in regard to the important questions about the origin of life and the reasons for our existence, large segments of society are still in the grip of superstition and prejudice, even though the study of more physical and benign subjects such as electricity and magnetism have long since broken free from the yokes of belief and wishful thinking.

Today we understand magnetism and electricity, and large segments of our world population benefit from a higher standard of living thanks to what the pioneers discovered. But those discoveries could be made only as the result of a scientifically rational approach to nature's secrets.

In 1609 Galileo Galilei turned his new telescope upward and found four pinpricks of light near Jupiter, which he realized were moons in orbit about that planet. That meant that one cherished belief, that the earth was the center of *all* motion, was plain wrong. Jupiter was the center of motion for its four moons, a shocking discovery for a geocentric clergy. Galileo also saw the phases of Venus, which proved that it orbited the sun, which proved that those two objects were not as perfect as everyone had believed for so long. For very good psychological reasons, Galileo's observations did not go over well with the Church. As the self-appointed and ultimate arbiter of truth, its role was implicitly challenged by Galileo's observations. Yet this man merely reported what he saw, and perhaps that is why the threat was so great. The scientific age would teach us to see more clearly, to perceive new truths that exist whether we want to see them or not. But, from the Church's point of view, if Galileo were allowed to report his "heresies," where would the rot stop?

In seeking answers to questions about the nature of life, we are still in the midst of an upheaval that dwarfs Galileo's struggle with the Church. One hundred fifty years after Darwin found evidence for the phenomenon of evolution, we are still witness to religious fundamentalist reaction to scientific explanations about the nature of life. Even in our enlightened age we find publishers quaking at the notion that local authorities will reject a biology textbook because it mentions evolution. Most scientists are polite about these issues, and go out of their way to accommodate beliefs held by millions of people. If we are to be honest in our pursuit of knowledge, we must surely recognize that there is a profound gulf between world views founded in belief and those built on experiment. Objectivity in the quest for answers requires that we set aside preconceptions and expectations. Like Faraday, we must recognize that we will only find answers by patiently attending nature's school.

The concept of evolution subsequently liberated the inquiring mind from the shackles of ancient beliefs about the nature of existence. As Richard Dawkins writes in his book *The Blind Watchmaker*,[5] "Our own existence once presented the greatest of all mysteries, but . . . it is a mystery no longer because it is solved." This is a remarkable thought, that the essential mystery is understood. But it is only solved in the eyes of those who look directly at nature and who are willing to observe the remarkable fruits of human curiosity. "I want to persuade the reader," Dawkins continues, "not just that the Darwinian world-view *happens* to be true, but that it is the only known theory that *could*, in principle, solve the mystery of our existence."[6]

This most fundamental of all mysteries, the nature of life, which for thousands of years has been dealt with in terms of superstitions—some of which evolved into religious dogma—remains in the murky depths of myth and metaphor. But ever since human curiosity was brought to bear on the question, in particular as demonstrated in recent progress in molecular biology, the blinders have been pulled back and scientists have found that a concept of an anthropomorphic, interfering God is no longer a sufficient or even a necessary cause for any of life's experiences. While individuals may feel or believe that God is involved, the essence of the scientific endeavor is to seek beyond feelings and beliefs to examine the nature of truth, the way things really are upon close and careful inspection and measurement.

Centuries may pass before the collective mind recovers from having dogmas and superstitions related to life destroyed. During such an era it may be impossible to recognize that much has changed. Yet, once the process of inquiry has begun, there is nothing that can stop human curiosity from finding the truth about nature. Once we have tasted of the apple of curiosity we are banished from the ignorance of Eden. Inevitably the process of questioning acquires a momentum of its own. It is an alarming fact that this exercise of seeking answers to satisfy our curiosity is not calculated to bring collective peace of mind, because what we discover about nature may sometimes be extremely disturbing.[7]

We asked questions to find answers. Persistent questioning has produced answers. In his quest Gilbert dispensed with wild beliefs to explain lodestone. But he had to deal with the subsequent uncertainty he felt by resorting to concepts such as "vapors" or magnetic "virtue" to account for magnetism. But those were only a temporary stopgap on the way to enlightenment, chimera that were wiped away once experimentation began. Will it be any different in other areas of human inquiry in the future?

NOTES

1. Thomas S. Kuhn, *The Structure of Scientific Revolutions.* (Chicago: University of Chicago Press, 1970), p. 162.

2. A host of examples of this phenomenon were given in the excellent television series and the books by James Burke, *Connections.* (Boston: Little, Brown, 1978), and *The Day the Universe Changed.* (Boston: Little, Brown, 1985).

3. Gerrit L. Verschuur, *The Invisible Universe Revealed: The Story of Radio Astronomy.* (New York: Springer-Verlag, 1987).

4. See, for example, H. C. Arp, G. Burbidge, F. Hoyle, J. V. Narliker, and N. C. Wickramasinghe, "The Extragalactic Universe: An Alternative View." *Nature* 346 (1990): 807.

5. Richard Dawkins, *The Blind Watchmaker.* (New York: W. W. Norton, 1986), p. ix.

6. Ibid., p. x.

7. Astronomical research has in recent years uncovered some very disturbing truths, not least of which is a list of violent events that have influenced evolution in the past and that threaten the continued existence of our civilization. Books about this subject have been written by the author, Gerrit L. Verschuur, *Cosmic Catastrophes.* (Reading: Addison-Wesley, 1978), and Clark Chapman and David Morrison, *Cosmic Catastrophes.* (New York: Plenum Press, 1989). The state of knowledge about the most immediate threat at the time of writing was reported by the author in an article entitled "This Target Earth," *Air and Space Magazine* (October/November 1991), which presents a report of the most serious danger, collision with a near-earth asteroid. The recognition of such dangers must inevitably shake humankind out of its comfortable ignorance of the true nature of the universe in which we find ourselves.

Index